SpringerBriefs in Computer Science

For further volumes:
http://www.springer.com/series/10028

Suguo Du • Haojin Zhu

Security Assessment in Vehicular Networks

Springer

Suguo Du
Antai College of Economics
and Management
Shanghai Jiao Tong University
Shanghai
People's Republic of China

Haojin Zhu
Department of Computer Science
and Engineering
Shanghai Jiao Tong University
Shanghai
People's Republic of China

ISSN 2191-5768   ISSN 2191-5776 (electronic)
ISBN 978-1-4614-9356-3   ISBN 978-1-4614-9357-0 (eBook)
DOI 10.1007/978-1-4614-9357-0
Springer New York Heidelberg Dordrecht London

Library of Congress Control Number: 2013951520

Printed on acid-free paper

Springer is part of Springer Science+Business Media (www.springer.com)

To my parents

Preface

Vehicular Networks, or Vehicular Ad-hoc Networks (VANETs), is regarded as a promising approach for future intelligent transportation system, and enables a wide range of safety and Infotainment applications. From the safety perspective, the introduction of VANETS greatly increases the safety of passengers by exchanging safety relevant information. From the Infotainment perspective, it exploits the Vehicle-to-vehicle communications and Vehicle-to-Road Side Unit communications to allow ubiquitous Internet Access, Video Streaming, Location-based Service, Content Distribution and Traffic Monitoring. The security and privacy is more than critical for the success of vehicular networks.

This book is designed to introduce some methods such as attack tree, attack-defense tree and attack-defense game from a system view for analyzing the vehicular network security and privacy problems. The existing research on VANETs security and privacy mainly focuses on the preventive techniques. From a system point of view, it lacks a comprehensive yet well-defined security evaluation to allow the system administrator to identify the most critical security threats and thus determine the appropriate defense strategy, which are more than important for the overall success of VANETs deployment. The existing risk analysis schemes include attack tree, attack graph or defense tree based solutions. However, there are several research challenges which make the existing security analysis solutions cannot work well for security and privacy evaluation in VANETs. Firstly, for VANETs security, the defense strategy is directly correlated to the attack strategy and vice versa, which means that the security evaluation should consider both of attack and the defense side rather than any single one. Secondly, most of the existing security solutions only consider how to prevent an attack while fail to consider the costs and gains of the attacker and the defender. In reality, a rational attacker or defender may try to maximize its attack or defense benefits instead of blindly launching an attack or adopting a countermeasure. Lastly, but no less importantly, how to model the mutual interaction between the attacker and defender remains a great challenge for VANETs security evaluation.

This book presents several novel approaches to model the interaction between the attacker and the defender and assess the security of the considered VANETs. The first security assessment approach presented in this book is based on attack tree security assessment model, which leverages tree based method to model and analysis the risk of the system and identify the possible attacking strategies the adversaries may launch. With the help of the attack tree model, it is convenient to analyze the capability of the attack source and estimate the degree or the impact a certain threat might bring to the system.

To further capture the interaction between the attacker and the defender, we further propose to utilize the attack-defense tree model to express the potential countermeasures which could be used to mitigate the system. The difference between an attack tree and an attack-defense tree is that the front only represents the attack strategies that attackers can launch, while the latter includes the set of countermeasures which can mitigate the possible damages produced by the attackers.

By considering rational participants that aim to maximize their payoff function, we propose a game-theoretic analysis approach to investigate the possible strategies that the security administrator and the attacker could adopt. On one side the VANETs security administrator wants to protect the security of the vehicular networks by adopting countermeasures to thwart the attacks; on the other side, the attacker wants to exploit the vulnerabilities and obtain some profit by attacking the vehicular networks. However, they cannot maximum their utility at the same time because one's action that aims to increase its own benefits will reduce its adversary's utility. Under this setting, we discuss the potential strategies of the defender and the attacker by modeling it as an attack-defense game. We then give a detailed analysis on its Nash Equilibrium.

Since many real world systems operate in multiple phases and, for mission success, all phases must be completed without failure. In practice, the attack strategy will evaluate from simple attack to more advanced yet complicated attacks along with the evolution of the defense strategy. Therefore, defending the attack succeeds if and only if the defense of all of phases succeed. We introduce a phased attack-defense game to model the interactions between the attacker and defender for VANET security assessment.

This book can be used for graduate students who interest in network security and privacy research area in their first year's literatures reviewing phase. This book can be a reference reading material for management science or computer science graduate students. The main mathematical prerequisite are the rudiments of Boolean algebra and game theory.

Finally we would like to thank Mr. Xiaolong Li and Mr. Junbo Du for their devotion to this work.

Shanghai, People's Republic of China   Suguo Du
August 2013   Haojin Zhu

Contents

Acronyms

AAA	Authentication, Authorization and Accounting
ALE	Annual Expected Loss
CA	Certificate Authority
CI	Cost of Investment
DoS	Denial-of-Service
EAP	Extensible Authentication Protocol
EBL	Extended Brake Lights
ECDSA	Elliptic Curve Digital Signature Algorithm
GI	Gain of Investment
ILD	Inductive Loop Detectors
MAC	Medium Access Control
OBU	On Board Unit
PKI	Public Key Infrastructure
RADIUS	Remote Authentication Dial In User Service
ROA	Return On Attack
ROI	Return on Investment
RM	Risk Mitigation
RSUs	Road Side Units
TLS	Transport Level Security
VANETs	Vehicular Ad-hoc Networks
V2I	Vehicle-to-Infrastructure
V2V	Vehicle-to-Vehicle

Chapter 1
Introduction to Vehicular Networks Security

Abstract In this Chapter, we give a brief introduction on vehicular networks, related applications, security requirements as well as the state-of-the-art of the vehicular security. We also introduce the system model and the organization of this book.

1.1 Vehicular Networks

With the advancement of wireless technology, vehicular communication networks, also known as vehicular ad-hoc networks (VANETs), are emerging as a promising approach to increase road safety and efficiency. In VANETs, vehicles are equipped with wireless communication, positioning and computing devices with a variety of vehicle applications enabled by communication between vehicles. For example, currently, the drivers can only see the brake light of vehicles ahead of them; and the brake light system can only demonstrate whether the vehicle is braking, but cannot indicate the level of deceleration. For example, when there is an emergency braking, drivers may not see the break lights of any other vehicles but the one in front of them, especially, when visibility is poor beyond the car in front of them (in fog), or in heavy traffic when everyone is so close or behind bigger vehicles like minivans, trucks, and Sport Utility Vehicles (SUVs). Under such a circumstance, rear-end collisions could happen with a much larger chance. On the contrary, if with VANETs, to countermeasure the situation, Vehicle-to-Vehicle communication can serve to extend the range of brake light signals for the drivers and as well indicate the level of deceleration (or referred to as Extended Brake lights (EBL)) [1]. Through the vehicle-to-vehicle (V2V) communications, the hard braking information of a vehicle is disseminated in a timely fashion so that the other vehicles can be alerted.

Besides these life-critical functionalities, another trend of functionalities which involves commercial related services is also envisioned in future VANETs [36–41]. These services including Internet access, real time traffic concerns, video streaming and content distribution, which are provided by some certain service provider, can be distributed initially from some wireless enabled roadside infrastructure, and then delivered through vehicle-to-infrastructure (V2I) and vehicle-to-vehicle (V2V) communications to the customers in vehicles. So-called infrastructure-based and services-oriented vehicular networks are expected to provide clear customer benefit and motivate service providers to invest on large-scale deployment of wireless infrastructures.

S. Du, H. Zhu, *Security Assessment in Vehicular Networks,* SpringerBriefs in Computer Science, DOI 10.1007/978-1-4614-9357-0_1, © The Author(s) 2013

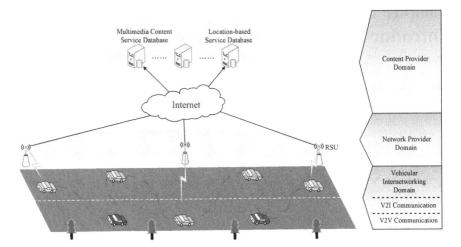

Fig. 1.1 Service-oriented Vehicular Networks: Components and Services

1.2 Applications in Emerging Vehicular Networks

Applications in service-oriented vehicular networks can provide road users with information, advertisements, and entertainment during their journey. In the literature [3], we summarize the existing applications and several potential applications that may be supported in service-oriented vehicular networks, which is shown in Fig. 1.1.

Internet Access Constant Internet access has become a daily demand for many of us and providing a Internet access is the basis for may other vehicular applications. To provide the Internet access, the mobile user should connect to the Internet gateway, which is served by the RSU. For example, Microsoft Corp.s MSN TV and KVH Industries, Inc. have introduced an automotive vehicle Internet access system called TracNet, which can bring the Internet service to any in-car video screens. It also turns the entire vehicle into an IEEE 802.11-based Wi-Fi hotspot, so passengers can use their wireless-enabled laptops to go online.

Broadcast Video Streaming The service provider, can broadcast the video streaming via roadside infrastructure (e.g., 802.11 access point) to vehicles driving through. Drivers or passengers could also enjoy watching live news or football match, while the video data is conveyed by either from RSU directly or from other relay vehicles.

Location-based Service There are many interesting location-based services that can be supported in VANETs. Finding restaurants, gas stations or available parking space in the nearby areas along the road are examples of such applications. Another example is carpooling service. Due to environmental concerns and rising gas prices, it is likely that carpooling will become an increasingly popular mode of transportation. For these kind of services, a service provider could maintain a server to collect and aggregate the service information, which can be reached by the passing vehicles via

RSUs. Each vehicle can issue a service finder request message that can be routed to the nearest RSU, and a service finder response message can be routed back to the vehicle.

Content Distribution Content distribution such as downloading digital map is another promising applications that may be supported by the VANETs. What differentiates content distribution from other infotainment services is the quantity of information, which is normally very large. Therefore, it is difficult for the vehicles to accomplish the file downloading with a limited number of transactions with RSUs or other vehicles and cooperative file distribution is inevitable in this case.

Traffic Monitoring As shown in [5, 26, 35], Departments of Transportation in the US must collect various types of data (e.g., average speed or traffic density) for traffic monitoring purposes. Traditionally, these important data are collected by technologies such as inductive loop detectors (ILD), video detection systems, acoustic tracking systems, or microwave radar sensors, which may be susceptible to failure or suffer from a high maintenance cost. On the other hand, cooperative data collection and dissemination in VANETs allow the traffic monitoring performed in a more cost-effective way [4]. Specifically, each vehicle collects its own or neighboring information (e.g., its current speed or neighboring traffic) and then transmits them to the remote roadside units (RSUs) via vehicle-to-vehicle(V2V) and vehicle-to-infrastructure(V2I) communications while the RSUs could be deployed at various points of interest along the roadway and can be used to collect data from locations up to tens of kilometers away [42].

From the underlying communication point of view, the aforementioned services can be classified into two categories: based on direct V2I communications or based on V2I and V2V hybrid communications. Even though direct V2I communication has the advantage on the transmission reliability and wireless bandwidth, it also suffers from limited transmission range. In the early stage of vehicular networks, it is not possible for the network/service operator to pursue a full coverage because of the high deployment and maintenance cost. Alternatively, the inter-vehicle transmission could broaden the effective range of service coverage. The emerging network coding techniques can be employed to improve the network throughput and increase the routing reliability. The advancement of wireless techniques have made the service-oriented vehicular networks possible in the near future.

1.3 Security Requirements for Emerging Vehicular Networks

Before discussing the details of the security requirements in service-oriented Vehicular Networks, let us give an example of a simple and typical service delivery communication in VANETs. Figure 1.2 shows that, a vehicle relies on V2I(shown in Fig. 1.2(a)) and V2V(shown in Fig. 1.2(b)) communication to get the Internet connectivity. The service requesting message generated and forwarded by the sending vehicle goes through intermediate nodes until arrive at a nearest RSU. Therefore, the security requirements can also be classified into V2V security and V2I security. In the following section, we will discuss the security requirements in vehicular networks.

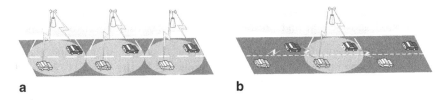

Fig. 1.2 V2I and V2V Communication in VANETs

Authentication In general, the authentication in VANETs includes two levels: authentication between the vehicles to provide link-to-link security, and authentication between the vehicle and RSU to provide end-to-end security.

- *V2V Authentication*: During the service delivery process, an attacker along the forwarding path can manipulate the transmitted data or try to launch Denial-of-Service (DoS) attacks on the network by jamming the radio channel at the link layer or at the routing layer by saturating the vehicles forwarding capacity. Other possible attacks on V2V communication include impersonation attack, where a malicious vehicle pretends to be another vehicle by forging one or more identities. Therefore, to provide link-to-link security along the data forwarding path, the V2V authentication is required. To achieve V2V authentication, we assume that each vehicle has a certified public/private key pair issued by an common offline security manager. These public/private key pairs are used to mutually authenticate the vehicles and establish the link-to-link security channel.
- *V2I Authentication*: To obtain the Internet connectivity and other services, RSU needs to authenticate the requesting vehicle to make sure that authentication, authorization and accounting (AAA) policy can be performed correctly. One promising way to achieve V2I authentication is adopting EAP/TLS based IEEE 802.1x authentication standard, with which the authentication request is issued by the vehicle and is sent through the RSU, until reaching a centralized AAA server maintained by the service provider (such as RADIUS) that can grant access to the user. The successful Vehicle/Service Provider mutual authentication authorizes each client to access the required services during his voyage.

Privacy Issue Privacy issues for service provisioning in VANETs regard primarily preserving the anonymity of a vehicle and/or the privacy of its location, which can be further classified into two privacy level: keeping privacy from other vehicles/external observer and from the RSU/Service Provider.

- *Keeping privacy from other vehicles*: In service oriented VANETs, the data/traffic may be forwarded under the help of the other vehicles. An adversary, which can be an malicious node along the forwarding path or a passive attacker, may try to link the vehicle to the drivers name, the license plate, speed, position, and traveling routes along with their relationships to compromise sender' privacy. A typical way to preserve users' privacy from the external observer is time-based pseudonyms, issued by a certificate authority and can be updated periodically.

- *Keeping privacy from RSU/Service Provider*: The user concerns are not only about the leaking his privacy to the public but also about the commercial misuse of their personal data by the service provider. For example, location privacy is among the most critical personal information and thus users may prefer to travel incognito. Therefore, in some cases, users may prefer to hide their identity from the service providers. Since the authentication with service provider normally involves the accounting and billing, the pseudonym based method can not work well in this case. In stead, anonymous authentication with the RSU or service provider can be achieved by using of the blind signature, which allows the message signed by the signer without revealing any information about the message.

1.4 Vehicular Security: State-of-the-Art

VANETs security and privacy is gaining an increased interest from both of industry and academia. In the past several years, there are quite a few studies on how to realize efficient data routing/forwarding in vehicular networks [2]. However, vehicular networks have brought new security challenges due to their mobile and infrastructureless nature. For example, the broadcast nature of the wireless medium allows an adversary to eavesdrop on the communications containing node identifiers, and to estimate the locations of the communicating nodes with an accuracy that is sufficient for tracking the nodes, which is referred to as privacy related threats. Further, a malicious vehicle could impersonate an legitimate user to disseminate bogus traffic information, which may mislead other vehicles and compromise the normal functionality of VANETs. Therefore, VANETs security and privacy is regarded as one of major challenges for vehicular communications. In this study, we mainly focus on how to protect the messages from being modified and how to preserve users' location privacy.

False message injection from outsider attacker is one of major security threats in VANETs. To provide the authentication and integrity checking for the broadcasted message, IEEE 1609.2 standard has proposed to have a Public Key Infrastructure (PKI) for key management. Each vehicle has a pair of ECDSA keys: a private signing key and a public verification key. The verification key is certified by a certificate authority (CA). Each sent message will append a signed signature to provide message authentication, which could prevent the outsider attackers from injecting bogus messages [7].

However, the insider false message injection attacker cannot be directly be addressed by the public key based solutions since the attackers could compromise a legitimate user and then exploit its private key to launch the attacks. To address the insider false message injection attack, there are two problems need to be addressed: how to detect a false message sent by a legitimate identity and how to revoke this legitimate but misbehaving node. For the first problem, one of the potential approaching is local voting approach which allows multiple vehicles to cross check a target message in VANETs [8]. For the second problem, VANETs could revoke a misbehaving node by revoke its public/private key pairs. That is, revocation decision making may be the result of a collaborative, systemic or a unilateral decision process [9]. In collaborative schemes, nodes accuse other nodes of misbehaving by casting negative votes against them. If a predetermined threshold of negative votes are cast, then the offending node is considered revoked [10–13]. By contrast, systemic revocation decision could be made by contacting a centralized CA. In the unilateral decision process, a notion of suicide has recently been extended for use in ad hoc networks where a node, upon detecting some malicious behavior, can instigate a suicide-bombing on a (perceived) malicious node. A node commits suicide by broadcasting a signed instruction to revoke both its own key and the key of the misbehaving node [14–18].

To protect privacy and prevent location tracking, a VANETs-enabled vehicle can obtain multiple certified key pairs with non-overlapping periods of validity and change its public key periodically (e.g., every five minutes) [19]. Note that the attacker could also launch the privacy related attack by tracking the long-term identifiers, such as MAC (Medium Access Control) addresses, IP address, or physical layer information. Therefore, the corresponding pseudonyms on different layers could be used to enhance the location privacy. To avoid the spatial-temporal correlation, the mix-zone based approach is introduced to enhance the location preserving by using the collaboration of multiple users. In addition, group signature based approach is another way to improve the location privacy [20].

Location Privacy Protection in location-based services is a long-standing topic and has received a lot of attentions in the last decades. The most popular approach to achieve location privacy in LBS is utilizing the obfuscation techniques to coarse the spatial or temporal granularity of the users' real locations [43, 44, 48–54]. A different approach to hide the users' location is based on mix zones. Mix zones are defined as the regions where users keep silent while changing their pseudonym together [45]. The third approach is to protect location privacy by adding dummy requests, which are issued by fake location and indistinguishable from real requests [46]. A recent work [47] proposes a game-theoretic framework that enables a designer to find the optimal LPPM for a given location-based service, ensuring a satisfactory service quality for the user.

In summary, there are quite a few threats and the corresponding protection solutions proposed for VANETs. The existing research on VANETs security and privacy mainly focuses on the preventive techniques. From a system point of view, it lacks a comprehensive yet well-defined security evaluation to allow the system administrator to identify the most critical security threats and thus determine the appropriate defense strategy, which are more than important for the overall success of VANETs deployment. The existing risk analysis schemes include attack tree, attack graph or defense tree based solutions. However, there are serval research challenges which make the existing security analysis solutions cannot work well for security and privacy evaluation in VANETs. Firstly, for VANETs security, the defense strategy is directly correlated to the attack strategy and vice versa, which means that the security evaluation should consider both of attack and the defense side rather than any single one. Secondly, most of the existing security solutions only consider how to prevent an attack while fail to consider the costs and gains of the attacker and the defender. In reality, a rational attacker or defender may try to maximize its attack or defense benefits in stead of blindly launching an attack or adopting a countermeasure. Lastly, but no less importantly, how to model the mutual interaction between the attacker and defender remains a great challenge for VANETs security evaluation.In the next section, we will give a more detailed threat analysis by using an attack-defense tree based model. The system model is as follows.

1.5 System Model

Communications in VANETs are divided into two parts: vehicle to infrastructure and vehicle to vehicle. The followings are some assumptions about the network [7]:

- Each vehicle has its own communication equipment OBU (On Board Unit), which enables the vehicles to communicate with others as well as the Road Side Units (RSUs).
- It is assumed that there is a trusted third party called Certificate Authority, like transportation authority within the network to take charge of the network's security and privacy issues. Each vehicle becomes a legitimate node of the network until it registers at CA.
- The CA disseminates each node with a single identity as well as a set of pseudonyms after it verified the validity of the node's identity.
- A node changes its pseudonym at certain intervals for the privacy preservation. Expired pseudonym is directly removed from the vehicle's storage media and CA is responsible for the issuance of new pseudonyms if a node uses up all of its pre-download pseudonyms.
- Each node automatically broadcasts its location, velocity and other special information to its neighbors at fixed intervals.
- Vehicles have enough power to install and run personal firewall or other antivirus software to protect it from malicious programs like worms and viruses spread among wireless network.

1.6 Summary

In this book, we introduce a new approach of attack tree method to make security assessment for vehicular networks. After this, we extend the attack tree model to attack-defense tree to analysis the interactions between attacker and defender. In particular, we consider both of the rational attacker and defender which decide whether to launch an attack or adopt a countermeasure based on its adversary's strategy to maximize its own attack and defense benefits. To achieve this goal, we firstly adopt the attack-defense tree to model the attacker's potential attack strategy and the defender's corresponding countermeasure. To take the attack and defense cost into consideration, we introduce two novel concepts: *Return On Attack*(ROA) and *Return on Investment*(ROI) to represent the potential gain from launching an attack or adopting a countermeasure. We further investigate the potential strategies of the defender and the attacker by using a game-theoretic analysis. In the newly defined attack-defense game, each rational participant may tend to get the maximum utility by maximizing *ROI* or *ROA*, which depends on the different utility attack/defense strategy and the associated attack/defense cost.

This book is organized as follows. In Chap. 1, we will introduce the emerging applications of vehicular networks, review the security requirements of vehicular networks, start-of-the-art of the standard and other related works. In Chap. 2, we will introduce an attack tree based security assessment approach. In Chap. 3, we extend the attack tree model to attack-defense tree model. In Chap. 4, a novel attack-defense game between the attacker and the defender is introduced and the Nash Equilibrium is analyzed. A case study is given to demonstrate the results of attack-defense game. In Chap. 5, we further introduce multiple phased attack-defense game model to capture the interaction between the attacker and the defender for multiple phased missions.

Chapter 2
Security Assessment via Attack Tree Model

Abstract Even though emerging as a promising approach to increase road safety, efficiency and convenience, Vehicular Ad hoc Networks (VANETs) pose many new research challenges, especially on the aspect of location privacy. The existing literatures focus on preventive techniques to achieve location privacy protection, however the location privacy risk assessment receives less attention. In this chapter, we introduce a novel risk assessment method to evaluate the security risk of VANETs privacy based on attack tree.

2.1 Introduction of Privacy Protecting in VANETs

Vehicular ad hoc networks [24] (or VANETs) are self-organized networks designed for communication among vehicles. In VANET, each vehicle is equipped with an On Board Unit, by which vehicles are able to communicate wireless with each other as well as Road Side Units. VANETs are expected to support a wide range of promising applications such as location based services. However, the broadcast nature of the wireless medium allows an adversary to eavesdrop on the communications containing node identifiers, and to estimate the locations of the communicating nodes with an accuracy that is sufficient for tracking the nodes. For example, in [25], it is reported that the adversary can even approximately derive the drivers family address and their workplaces with the collection of location traces every day. Due to these reasons, location privacy threat [26] has been well recognized as one of the major security threats for VANETs and has attracted a lot of interest recently.

The existing research on VANET privacy mainly focuses on the preventive techniques including: pseudonym [27] based approaches, group signature [28] based techniques or mix-zone [26, 29, 30] based approach. The basic idea of the preventive techniques is introducing the privacy protection techniques to prevent the compromise of user location privacy. However, the preventive schemes may face the challenges that it can only address the expected security vulnerabilities while have nothing to do with the unexpected privacy threats. Furthermore, from a system point of view, a comprehensive yet well-defined security evaluation enables the system administrator to identify the most critical security threats and attack strategies, which are more than important for the overall success of VANET deployment. Even though some reported studies also investigate the possible security vulnerabilities, they fail to give a quantity risk analysis from a system point of view.

S. Du, H. Zhu, *Security Assessment in Vehicular Networks*, SpringerBriefs in Computer Science, DOI 10.1007/978-1-4614-9357-0_2, © The Author(s) 2013

In this study, we propose a novel risk assessment approach for location privacy preserving in VANETs based on the attack tree based approach. Attack tree based risk analysis leverages tree based method to model and analysis the risk of the system and identify the possible attacking strategies the adversaries may launch. With the help of the attack tree model, it is convenient to analyze the capability of the attack source and estimate the degree or the impact a certain threat might bring to the system. Due to these features, in this chapter, we take advantage of attack tree based approach to identify the possible threats. And we further calculate the total probability of reaching attack goal on the basis of the attack tree. According to the quantitative result, the decision maker of the system can decide which protection measure should be adopted.

The reminders of this chapter are organized as follows: in Sect. 2.2, we introduce the attack tree method and present how to build the attack tree. In Sect. 2.3, we assign values to leaf nodes and calculate the systems risk. In Sect. 2.4, to estimate the most likely way an attacker may choose, we carry out an analysis of attack scenarios on the basis of attack tree. At last, in Sect. 2.5, we summarize the conclusion .

2.2   Attack Tree Model for VENET Privacy

An attack tree [31] can be simply described as an analytical technique, whereby an undesired state of the system is specified, and the system is then analyzed in the context of its environment and operation to find all credible ways in which the undesired event can occur [32]. There are two basic types of attack tree gates: the OR-gate and the AND-gate. The OR-gate is used to show that the output event occurs only if one or more of the input events occur. The AND-gate shows that the output attack occurs only if all the input attacks occur. To analyze the system, we select a particular event of the system as an attacker's goal, and then determine the immediate, necessary, and sufficient causes for the occurrence of this goal. It should be noted that these are not the basic causes of the goal but the immediate causes for the event. These immediate, necessary, and sufficient causes of the goal are now treated as sub-goals and we proceed to determine their immediate, necessary, and sufficient causes. In this way we proceed down the tree continually, until ultimately we reach the limit of resolution of our tree, that is leaf node (atomic attack) of an attack tree.

In VANET system, we choose "leakage of location privacy", which is denoted by G, as the attack goal. Now we proceed with a step-by-step analysis of the attack goal. The intermediate causes of the goal are: direct communication, eavesdropping, stealing and illegal disclosure, which are respectively marked with M_1, M_2, M_3, and M_4. The mission objective can be achieved if any of the four components is reached. Now we identify the four intermediate causes as sub-goals, and it is necessary to determine their immediate cause or causes separately.

There are two possibilities for direct communication (M_1).

"Inquiry (X_1)": an attacker communicates with a target node with real identity of itself, and then inquiries the node's location privacy. This way applies only when a node has low sensitivity on its privacy.

"Cheating (M_5)": an attacker impersonates as someone else that the target node trusts, and gets privacy of the target node by communicating with it. Since nodes which have close relationship with a target node such as friends, colleagues or service providers are supposed to be reliable for most nodes, this kind of attack succeeds with higher opportunity.

Therefore, the sub-goal "direct communication (M_1)" can arise from two events, "inquiry (X_1)" or "cheating (M_5)". Now we are ready to seek out the immediate causes for the new sub-goal "cheating (M_5)", which appears as intersection of two events: "finding vulnerabilities in the systems authentication mechanism X_2" and "making fake identity X_3".

We now continue the analysis by focusing our attention on event "eavesdropping (M_2)". To get a node's privacy by "eavesdropping (M_2)", it is necessary to carry out "physical layer eavesdropping (M_6)", or "MAC layer eavesdropping (M_7)", or "application layer eavesdropping (M_8)". We identified M_6, M_7, and M_8 as intermediates which will be analyzed as below.

Physical layer eavesdropping (M_6): This mission can be achieved if two tasks were accomplished in a row: "dismantle wiretap-proof device X_4" and "installing wiretap device M_{11}". There are three ways for an attacker to successfully "installing wiretap device M_{11}": "being a service provider for cars X_5", or "disrupting the cars anti-theft system X_6", or "making use of the owners carelessness X_7".

MAC layer eavesdropping (M_7): The attack "MAC layer eavesdropping (M_7)" can be launched through two joint atomic attacks: "protocol vulnerability analysis X_8" and "resetting its own configuration X_9".

Application layer eavesdropping (M_8): In application layer, an attacker can choose from two aspects to eavesdrop: "eavesdropping pseudonyms M_{12}" or "running eavesdropping software M_{13}". Each of the two is further decomposed into sub-components. For "eavesdropping pseudonyms M_{12}", it is essential for an attacker to "obtaining signal receiver X_{10}" and "analyzing the adopted pseudonym mechanisms weaknesses X_{11}". While for "running eavesdropping software M_{13}", after "breaking the networks firewall X_{12}" an attacker needs to "being familiar with wireless networks weak security feature X_{13}" and then compromise the target node by planting eavesdropping program.

Sub-components of "stealing M_3" include "physical theft M_9" and "malicious node theft M_{10}". These two sub-components are represented by a logical OR relationship in the attack tree construction, for any occurrence of the two possible events which may result in the happening of "stealing M_3". In the attack "physical theft M_9", three steps must be taken: "stealing the car M_{14}", "disrupting the function of removing data from remote control X_{15}" and "deciphering encrypted file X_{14}". When an attacker steals a car, he can adopt the same approaches as in "installing wiretap device M_{11}". "malicious node theft M_{10}" refers to an attack using malicious program to steal privacy stored in the vehicles storage media. It requires an attacker to "deciphering encrypted file X_{14}" as well as "breaking the networks firewall X_{12}".

As far as the last intermediate cause of the attack goal "illegal disclosure M_4" is concerned, it is decomposed into a logical OR relationship of two atomic attacks: "purchasing privacy information from third party X_{16}" or "leakage from official department X_{17}". Third parties such as location service providers, employees collect large information about nodes privacy, so it is possible that these trusted third parties might sell their collected privacy for commercial profit. For the convenience of management, official management department could also leak nodes privacy accidentally.

According to the analysis above, we built the attack tree model which is presented in Fig. 2.1. Notations of gates and leaf nodes in the attack tree are listed in Table 2.1.

2.3   Risk Assessment

For the limitation of resources, an attacker has to take into account of many aspects including the possibility to succeed, attack cost, technique difficulty, risk of being detected and so on. In this work, we calculate the total probability of reaching the attack goal by assigning leaf nodes three attributes: attack cost, technical difficulty and discovering difficulty, which are denoted as c_L, d_L and s_L respectively. The grade level standards are given in Table 2.2. The values of c_L, d_L and s_L depend on the following rules.

An attacker could launch an attack on any layer of the system. The higher the attacked layer is, the more difficult for the attacker, and the lower the probability of success. In this condition, we can sort those layers according to the difficulty of compromise in the following order: application layer > transmission layer > routing layer > MAC layer > physical layer.

Cost of physical vehicle and labor assistant is higher compared with that of analyzing protocol or mechanisms vulnerabilities. Powerful nodes, for instance traffic offices or service providers, are capable of taking more protecting behaviors, so these nodes are more difficult to be compromised than common nodes.

The multi-attribute utility theory [34] is adopted to transfer these three attributes into attackers utility value, which is the occurrence probability of a leaf node. The following is a formula we applied to calculate the utility of each leaf node:

$$P_L = w_1 \times u(c_L) + w_2 \times u(d_L) + w_3 \times u(s_L)$$

Where $u(c_L)$, $u(d_L)$, $u(s_L)$ represent the utility functions of c_L, d_L and s_L, and their values fall into the interval of [0, 1]; w_1, w_2, w_3 are the weights of the utilities, where $w_1 + w_2 + w_3 = 1$.

For the convenience of calculation, we define that $w_1 = w_2 = w_3 = 1/3$. The specific assignment of each nodes attribute requires knowledge of implementation details of the system in question including protocols, hardware, operating system as well as attack software and tools. Since the main focus of this chapter is proposing a new evaluation method, we assign the value to each leaf node, which is listed in Table 2.3. In order to get the occurrence probability of each leaf node, we also

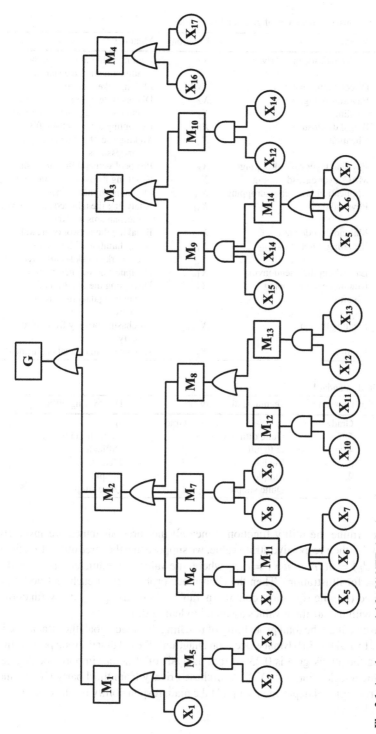

Fig. 2.1 Attack tree model for VANETs location privacy

Table 2.1 Notations and meanings of gates and leaf nodes

Notations	Meaning	Notations	Meaning
G	Leakage of location privacy	X_2	Finding vulnerabilities in the authentication mechanism
M_1	Direct communication	X_3	Making fake identity
M_2	Eavesdropping	X_4	Dismantling wiretap-proof device
M_3	Stealing	X_5	Being a service provider for cars
M_4	Illegal disclosure	X_6	Disrupting a cars anti-theft system
M_5	Cheating	X_7	Making use of the owners carelessness
M_6	Physical layer eavesdropping	X_8	Protocol vulnerability analysis
M_7	MAC layer eavesdropping	X_9	Resetting its own configuration
M_8	Application layer eavesdropping	X_{10}	Obtaining signal receiver
M_9	Physical theft	X_{11}	Analyzing the adopted pseudonym mechanisms weakness
M_{10}	Malicious code theft	X_{12}	Breaking the networks firewall
M_{11}	Installing wiretap tool	X_{13}	Being familiar with wireless networks weak security feature
M_{12}	Eavesdropping pseudonyms	X_{14}	Deciphering the encrypted file
M_{13}	Running eavesdropping software	X_{15}	Disrupting the function of removing data from remote control
M_{14}	Stealing the car	X_{16}	Purchasing privacy from third party
X_1	Inquiry	X_{17}	Leakage from official department

Table 2.2 Grade standard

Attack cost/Ten Thousands		Technical difficulty		Discovering difficulty	
c_L	Grade	d_L	Grade	s_L	Grade
> 10	5	Quite difficult	5	Quite difficult	1
6–10	4	Difficult	4	Difficult	2
3–6	3	Mediate	3	Mediate	3
0.5–3	2	Simple	2	Simple	4
< 0.5	1	Quite simple	1	Quite simple	5

need to determine the utility function. Since all the three attributes are inversely proportional to their respective utility value, we suppose that the three utility functions $u(c_L) = u(d_L) = u(s_L) = u(x) = c/x$ (where, the value of parameter c is set as 0.2 in this book for illustration). Then the occurrence probability of each leaf node can be calculated (see the rightmost column in Table 2.3) by using the utility functions combined with the attribute values assigned to leaf nodes.

So as to calculate the total probability of reaching the attack goal, the attack tree is transferred to a BDD [33] (Binary Decision Diagram). We get that the total probability of reaching the attack goal is 0.239. From analysis of structure importance degree, three atomic attacks "inquiry (X_1)", "purchase privacy from third party (X_{16})" and "leakage from official department (X_{17})" take main responsibility for the occurrence of the attack goal.

Table 2.3 Attribute calues for leaf nodes

Leaf node	Attribute			Occurrence probability
	Attack cost	Technical difficulty	Discovering difficulty	
X_1	3	2	5	0.069
X_2	3	4	2	0.072
X_3	2	1	4	0.117
X_4	1	2	2	0.133
X_5	5	1	5	0.093
X_6	2	3	4	0.072
X_7	1	2	3	0.122
X_8	3	3	1	0.111
X_9	4	2	1	0.117
X_{10}	2	4	2	0.083
X_{11}	3	4	3	0.061
X_{12}	2	3	2	0.089
X_{13}	2	5	1	0.113
X_{14}	1	2	4	0.117
X_{15}	4	5	5	0.043
X_{16}	4	4	3	0.056
X_{17}	5	2	5	0.060

2.4 Attack Scenarios

An attack scenario [34] is a set of leaf nodes, in which only the occurrence of all the leaf nodes could reach the attack goal, that is to say the goal will not be realized if one of the leaf nodes does not occur. An attack scenario is the real attack way that an attacker considers. Once the attack scenarios have been known, we could calculate their probabilities of occurrence, and then compare them to find out the attack scenario that the malicious may launch most likely. Suppose an attack scenario is denoted as:

$$S_i = (X_{i_1}, X_{i_2}, \ldots , X_{i_n})$$

Then the probability of an attack scenario is:

$$P(S_i) = P(X_{i_1}) \times P(X_{i_2}) \times \ldots \times P(X_{i_n}) \tag{2.1}$$

We adopt Boolean algebra method to get all the attack scenarios for our attack tree. It can be seen that there are fourteen attack scenarios to achieve the attack goal, and they respectively are $\{X_1\}, \{X_2, X_3\}, \{X_4, X_5\}, \{X_4, X_6\}, \{X_4, X_7\}, \{X_8, X_9\}, \{X_{10}, X_{11}\}, \{X_{12}, X_{13}\}, \{X_5, X_{14}, X_{15}\}, \{X_6, X_{14}, X_{15}\}, \{X_7, X_{14}, X_{15}\}, \{X_{12}, X_{14}\}, \{X_{16}\}, \{X_{17}\}$. In the first scenario $\{X_1\}$, it means an attacker can get the targets privacy by just launching the atomic attack X_1, while in the second scenario $\{X_2, X_3\}$, an attacker requires to compromise the systems authentication mechanism X_2 as well as making fake identity X_3. From the Eq. 2.1, we calculate probabilities of occurrence for attack sequences which are listed in Table 2.4.

Table 2.4 Probabilities of attack sequences

Attack sequence	Probability
S_1	0.0690
S_2	0.0084
S_3	0.0123
S_4	0.0096
S_5	0.0162
S_6	0.0130
S_7	0.0051
S_8	0.0101
S_9	0.0005
S_{10}	0.0004
S_{11}	0.0006
S_{12}	0.0104
S_{13}	0.0560
S_{14}	0.0600

From the Table 2.4, we find that the attack scenario S_1 is the most likely happened attack method, so to protect the system from attack, it is necessary to first keep close eyes on it and take correspondent location protection measures.

2.5 Summary

Location privacy in VANETs is getting more concerns of the world. However, the study on the risk assessment of location privacy in the system has received less attention. In this chapter, we present an attack tree based security assessment methodology to quantify the risk of location privacy in VANETs from the systems perspective. We further build an attack tree with the leakage of location privacy information as attack goal. Also the total possibility of reaching attack goal is calculated on the basis of the attack tree. At last, according to the attack scenario analysis, we find out the most likely path that an attacker may use. In the next chapter, we will introduce an attack-defense tree method for VANETs security and privacy assessment, by which the system's defense strategies can be analyzed.

Chapter 3
Attack-Defense Tree Based Security Assessment

Abstract In this Chapter, we present an attack-defense tree model for VANETs security and privacy problems, which considered both attacker's and defender's strategies. We also introduce two concepts of Return on Attack and Return on Investment to represent the potential gain from launching an attack or adopting a countermeasure. Both the attack-defense tree and the introduced concepts are preparation work for the attack-defense game analysis of later work in Chap. 4.

3.1 Introduction to Attack-Defense Tree Model

In the previous chapter, we adopt attack tree approach to model the behavior of the attackers in VANETs system and the effect of exploits. In general, attack trees offer a goal-oriented perspective that facilitates the expression of multi-stage attacks [17]. Attack trees in their simplest form assert subgoals for achieving the goal set forth by an attack node. Attack nodes can be grouped into AND or OR sequences to capture conjunctive and disjunctive attack conditions, respectively. Nodes can be weighted to reflect the likelihood of successfully mounting an attack. However, the attack tree in last chapter does not consider the defender's countermeasures or actions. We further propose to utilize the the attack-defense tree model to express the potential countermeasures which could be used to mitigate the system. The difference between an attack tree and an attack-defense tree is that the front only represents the attack strategies that attackers can launch, while the latter includes the set of countermeasures which can mitigate the possible damages produced by the attackers [21].

Figure 3.1 illustrates the structure of an attack-defense tree. There are two parts in an attack-defense tree. The square nodes represent the attack goals or actions which form the attack-tree part; the circle nodes represent the corresponding countermeasures of each attack goal or action. The top of the attack tree is associated with the asset of the system under consideration, which represents the attacker's final objective. The atomic attack (or leaf node)in the attack tree can lead the attacker to (partially) damage the asset by exploiting a single vulnerability. The sub-goal nodes (or Non-leaf nodes)can be of two different types under two kinds of gates: or-nodes (under or-gates) and and-nodes (under and-gates). Sub-goals associated with or-nodes are achieved as long as any of its child nodes is achieved, while and-nodes represent the sub-goals which require all of its child nodes to be completed.

S. Du, H. Zhu, *Security Assessment in Vehicular Networks,* SpringerBriefs in Computer Science, DOI 10.1007/978-1-4614-9357-0_3, © The Author(s) 2013

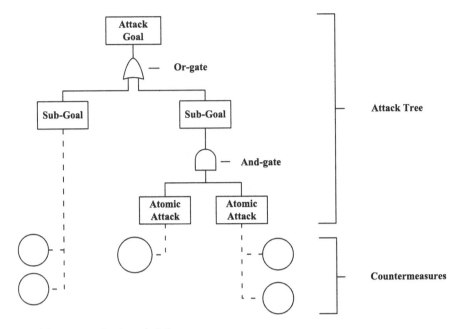

Fig. 3.1 An example of attack-defense tree

3.2 Building Attack-Defense Tree for VANETs Security

In this section we will build the attack-defense tree model for VANETs security analysis. In VANETs system, we set *Compromise the Security or Privacy* of VANETs as attack tree root, which is denoted as *A*. Two sub-goals of *A* are *Compromise Communication Security* and *Compromise Location Privacy*, which are denoted as *AS* and *AP*, respectively. In other words, an attacker could achieve the attack objective *A* by compromising the communication security (or the left sub-goal) or location privacy of the vehicles (or the right sub-goal). The attack-defense tree for VANETs security and privacy is shown in Fig. 3.2.

In the left sub-tree, there are three possible ways to achieve sub-goal *AS*, *Outsider False Data Injection* (AS_1), *Insider False Data Injection* (AS_2) and *Denial of Service* (AS_3). To achieve AS_1, the attacker must take two actions: to impersonate a legitimate vehicle(*Masquerading*) denoted by AS_{11} and to disseminate *Bogus Information* denoted by AS_{12} (e.g. misleading traffic information). The countermeasure corresponding to this attack is *PKI & Signature* (C_1), which means that the security authority could prevent the outsider attackers from distributing unauthorized messages by requiring each vehicle to provide the public key signature for each sending message. Similarly, to obtain the sub-goal "insider false data injection (AS_2)", it is necessary for the attacker to perform two attack steps *Stealing Identity* (AS_{21}) and *Bogus Information* (AS_{22}), in which an attacker could compromise a legitimate node's secret key and disseminate the unauthentic information by using the compromised

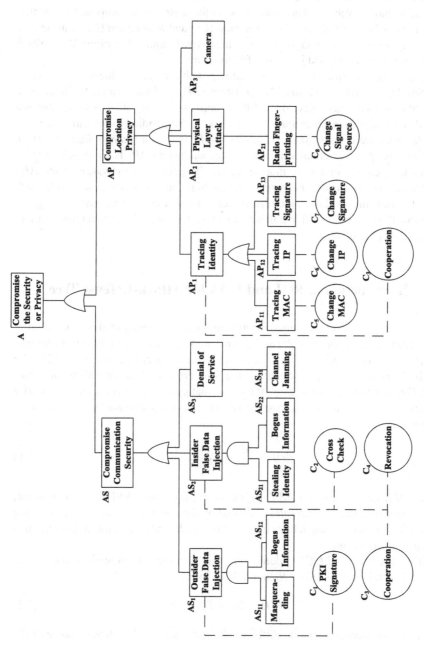

Fig. 3.2 The attack-defense tree model for VANETs security and privacy

secret key. To thwart insider attack AS_2, the security authority could cross check the information under the collaboration of multiple nodes and revoke the public key of a misbehaving vehicle. Therefore, these defense strategies compose the counter-measures of *Cross Check* (C_2), *Cooperation* (C_3) and *Revocation* (C_4). Further, the *Denial of Service Attack* (AS_3) could be achieved by *Channel Jamming* (AS_{31}), which cannot be easily addressed in a cost-effective way.

In the right sub-tree, there are also three possible ways to achieve sub-goal AP, *Tracing Identity* (AP_1), *Physical Layer Attack* (AP_2) and *Camera* (AP_3). To achieve AP_1, the attacker may select one of three actions: *Tracing MAC* (AP_{11}), *Tracing IP* (AP_{12}) and *Tracing Signature* (AP_{13}). The corresponding countermeasures for the defender are *Change MAC* (C_5), *Change IP* (C_6) and *Change Signature* (C_7), respectively. Another countermeasure corresponding to AP_1 is *Cooperation* (C_3), with which C_5, C_6 or C_7 are more effective. To achieve the physical attack AP_2, the attack must use *Radio Fingerprinting* (AP_{21}). *Change Signal Source* (C_8) such as radio transmitters that randomize fingerprints is a countermeasure to AP_{21}. In addition, the *Camera* (AP_3) is a thorny attack so that there is no countermeasure to mitigate AP_3 effectively.

3.3 Introduction of ROI and ROA for Attack-Defense Tree

In this work, we consider economic factors in VANETs security analysis by introduc-ing Return on Investment (*ROI*) and Return on Attack (*ROA*). In the attack-defense tree, there are two types of costs: cost of attack and security investment cost [21]. We firstly define *Return on Investment* for the countermeasures of defenders. In partic-ular, *ROI* is introduced to measure the return that a defender expects from a security or privacy investment over the costs he sustains for countermeasures. It is defined as follows:

$$ROI = \frac{ALE \times RM - CI}{CI} \qquad (3.1)$$

where *ALE* denotes the *Annual Expected Loss* caused by VANETs security threat; *RM* represents the *Risk Mitigation* induced by the countermeasure; *CI* denotes the *Cost of Investment* which defines the cost that the defender pays for implementing a given countermeasure.

If we use $R_D = ALE/CI$ to denote the gain-cost ratio for defenders, the *ROI* in Eq. (3.1) can be expressed by R_D and *RM* as follows:

$$ROI = R_D \times RM - 1 \qquad (3.2)$$

We will give more analysis on different alternatives of *ROI* with the change of R_D and *RM* in case study section. We further define the *Return On Attack* (*ROA*), which is used to measure the gain that an attacker expects from a successful attack over

Table 3.1 Evaluation of *ROI*

Attack	ALE	Countermeasures	RM	CI	ROI
AS_1	4	C_1	1	2	1
		$C_2, \cdots, C_8$	0	–	−1
AS_2	10	C_2	0.25	4	−0.375
		C_3	0.5	8	−0.375
		C_4	0.25	4	−0.375
		$C_1, C_5, \cdots, C_8$	0	–	−1
AS_3	8	$C_1, \cdots, C_8$	0	–	−1
AP_{11}	4	C_3	0.75	8	−0.625
		C_5	0.25	4	−0.75
		$C_1, C_2, C_4, C_6, C_7, C_8$	0	–	−1
AP_{12}	4	C_3	0.75	8	−0.625
		C_6	0.25	4	−0.75
		$C_1, C_2, C_4, C_5, C_7, C_8$	0	–	−1
AP_{13}	4	C_3	0.75	8	−0.625
		C_7	0.25	4	−0.75
		$C_1, C_2, C_4, C_5, C_6, C_8$	0	–	−1
AP_2	6	C_8	1	6	0
		$C_1, \cdots, C_7$	0	–	−1
AP_3	8	$C_1, \cdots, C_8$	0	–	−1

the losses that he sustains due to the adoption of security or privacy measures by his target. *ROA* is defined as follows:

$$ROA = \frac{GI \times (1 - RM) - (Cost_A + Cost_{AC})}{Cost_A + Cost_{AC}} \tag{3.3}$$

where *GI* represents the expected gain from a successful attack on the specified target; $Cost_A$ is the cost sustained by the attacker to launch an attack, and $Cost_{AC}$ represents the additional cost brought by the countermeasure *C* adopted by the defender to mitigate the attack.

Similar with the definition of R_D, we let $R_A = GI/(Cost_A + Cost_{AC})$ denote the gain-cost ratio for attackers. Consequently *ROA* in Eq. (3.3) is reduced as follows:

$$ROA = R_A \times (1 - RM) - 1 \tag{3.4}$$

Since *ROA* is a function of R_A and *RM*, we will discuss the different impacts on *ROA* caused by R_A and *RM* in later case study section.

From the definitions of *ROI* and *ROA*, the values of them are relative and they represent the return of the costs. In our case, according to the attack-defense tree and the possible difficulties of acting countermeasures and attacks, we use a set of levels $0, 2, 4, 6, 8, 10$ as the specific values of *ALE, CI, GI, $Cost_A$* and $Cost_{AC}$, by which obtaining the final values of the gain and cost is possible. Since *RM* is a risk measurement, we will choose $0, 0.25, 0.5, 0.75, 1$ as the specific values of *RM* to represent the effectiveness of each countermeasure. A higher value of *RM* indicates a more effective countermeasure. Thus, we can evaluate the *ROI* and *ROA* for all the countermeasures according to the data in Tables 3.1 and 3.2 respectively [22].

Table 3.2 Evaluation of ROA

Attack	GI	$Cost_a$	Countermeasures	$Cost_{ac}$	ROA
AS_1	4	4	C_1	0	-1
			$C_2, \cdots, C_8$	0	0
AS_2	4	10	C_2	4	-0.786
			C_3	8	-0.889
			C_4	4	-0.786
			$C_1, C_5, \cdots, C_8$	0	-0.6
AS_3	2	6	$C_1, \cdots, C_8$	0	-0.667
AP_{11}	6	4	C_3	8	-0.875
			C_5	6	-0.55
			$C_1, C_2, C_4, C_6, C_7, C_8$	0	0.5
AP_{12}	6	4	C_3	8	-0.875
			C_6	6	-0.55
			$C_1, C_2, C_4, C_5, C_7, C_8$	0	0.5
AP_{13}	6	4	C_3	8	-0.875
			C_7	6	-0.55
			$C_1, C_2, C_4, C_5, C_6, C_8$	0	0.5
AP_2	8	6	C_8	0	-1
			$C_1, \cdots, C_7$	0	0.333
AP_3	8	10	$C_1, \cdots, C_8$	0	-0.2

In Table 3.1, it should be noticed that the reason of choosing the subgoals AS_1, AS_2 instead of atomic attacks AS_{11}, AS_{12}, AS_{21}, AS_{22} for attackers in the Attack column which is related to the logical gate type of the attack subgoals. Due to the 'AND' logical type of AS_1, AS_2 shown in Fig. 3.2, the attacker won't achieve any gain by only taking one single attack of AS_{11}, AS_{12}, AS_{21}, AS_{22}.

In the RM column, zero values indicate that the countermeasure cannot mitigate the attack. For example, countermeasure $C_2, \ldots, C_8$ cannot mitigate attack AS_1, the corresponding RM is 0. The dash "-" in the column of CI means that the CI of the countermeasure do not impact the result of the corresponding ROI in Table 3.1. In addition, when a countermeasure cannot mitigate an attack ($RM = 0$), in this case, $ROI = -1$ in Table 3.1 and $ROA = (GI - Cost_A)/Cost_A$ in Table 3.2.

From the above, we can obtain the values of ROI and ROA between each pair of countermeasure and attack, which constitute the utility matrix for the attack-defense game in the next chapter.

3.4 Summary

In this Chapter, we use an attack-defense tree to model the actions of the attacker and the defender in VANETs security and privacy problems. Further, the concepts of ROA and ROI are introduced for evaluating the gains of attacker and defender. Based on the attack-defense tree, we will present an attack-defense game for VANETs security and privacy in next Chapter.

Chapter 4
A VANETs Attack-Defense Game

Abstract The existing risk analysis solutions may not work well to evaluate the security threats in vehicular networks due to the lack of considering the attack and defense costs and gains, and thus cannot appropriately model the mutual interaction between the attacker and defender. In this study, we consider both of the rational attacker and defender who decide whether to launch an attack or adopt a counter-measure based on its adversary's strategy to maximize its own attack and defense benefits. We investigate the potential strategies of the defender and the attacker by modeling it as an attack-defense game. The equilibriums of the game are given and the rationality of the proposed game-theoretical model is illustrated by a detailed case study.

4.1 Game Model

In the previous chapter, we have introduced *Return on Attack (ROA)* to measure the effectiveness of attacks in terms of attack cost and *Return on Investment (ROI)* to evaluate the investment on a security countermeasure with regard to a specific attack. On one side the VANETs security administrator wants to protect the security of the vehicular networks by adopting countermeasures to thwart the attacks; on the other side, the attacker wants to exploit the vulnerabilities and obtain some profit by attacking the vehicular networks. By using *ROI* and *ROA* to represent the utility of the defender and the attacker, both of the defender and the attacker may tend to get the maximum utility by maximizing *ROI* or *ROA*, respectively. However, they cannot maximum their utility at the same time because one's action that aims to increase its own benefits will reduce its adversary's utility. Therefore, in this chapter, we investigate the possible strategies of the security administrator and of the attacker by using a game-theoretic analysis. We consider rational participants that maximize their payoff function, which depends on the different utility attack/defense strategy and the associated attack/defense cost [23].

In this section we analyze the possible strategies of the system defender and of the attacker by using an attack-defense game model for VANETs security and privacy. Game theory allows modeling situations of conflict and, hence, predicting the behavior of the participants. We model the attack-defense game as a *static game* in which each participant take action when another's action is unknown. It is reasonable since in our VANETs security attack-defense system, both defender and attacker

S. Du, H. Zhu, *Security Assessment in Vehicular Networks,* SpringerBriefs in Computer Science, DOI 10.1007/978-1-4614-9357-0_4, © The Author(s) 2013

Table 4.1 The Utility Matrix of Attack-Defense Game

		Attack		
		A_1	$\cdots$	A_8
	C_1	$ROI(C_1,A_1),ROA(C_1,A_1)$	$\cdots$	$ROI(C_1,A_8),ROA(C_1,A_8)$
Defense	$\vdots$	$\vdots$	$\ddots$	$\vdots$
	C_8	$ROI(C_8,A_1),ROA(C_8,A_1)$	$\cdots$	$ROI(C_8,A_8),ROA(C_8,A_8)$

don't know each other's strategies when they take actions. We also suppose that the participants are rational in our model. This model assumption keeps our analysis tractable while solving the Nash Equilibrium solution of the game. The game $\mathcal{G}$ is defined as a triplet $(\mathcal{O};\mathcal{S};\mathcal{U})$, where $\mathcal{O}$ is a set of players, $\mathcal{S}$ is a set of strategies and $\mathcal{U}$ is a set of payoff functions.

- **Players**: The set of $\mathcal{O} = \{O_i, i = 1, 2\}$ includes a defender O_1 and an attacker O_2, noted as Def and Att respectively. Each player has no idea about which action his adversary has chosen (e.g. as soon as the attacker has decided to perform the insider false data injection (AS_2), the defender can't receive any information about it so that the game is static.).
- **Strategy**: Each player has a set of strategies $S_k(k = 1, 2$: all countermeasures $C_i \in S_1$ and all attacks $A_j \in S_2)$. According to our attack-defense tree model for VANETs security and privacy, the countermeasures which the defender can select are $\{C_i | i = 1, \ldots, 8\}$; the attacks which the attacker can select are $\{A_j | j = 1, \ldots, 8\}$, or $\{AS_1, AS_2, AS_3, AP_{11}, AP_{12}, AP_{13}, AP_2, AP_3\}$ respectively.
- **Payoff function**: The utility functions(or payoff) are defined as: $u_1(C_i, A_j) = ROI(C_i, A_j)$; $u_2(C_i, A_j) = ROA(C_i, A_j)$.

We show the utility matrix of the attack-def ense game in Table 4.1. We suppose the players know the utility (payoff) functions with each other completely, thus our game is a game of *complete information*. It is noticed that knowing the adversary's utility doesn't give a clue to the adversary's action strategies. For example, the defender knows that the cost or the gain to launch an location privacy related attack. However, he has no idea about which specific attack action (e.g., tracing the identity, physical layer tracing, or even use camera) will be adopted by the attacker.

4.2   Equilibrium Concepts

In this section, we introduce a few game-theoretic concepts that will help us get an insight into the strategies of participants. In our complete information attack-defense game, a *pure strategy* for two players is (C_m, A_n), $C_m \in S_1$ and $A_n \in S_2$, which means that under certain conditions the strategies of the attacker and of the defender converges to a pair of best action profile. This is to say that the defender cannot do better by choosing an action different from C_m, given that the attacker adopt A_n, and vice versa. In this case we say that our attack-defense game admits a Nash Equilibrium.

Definition 1 (**Nash Equilibrium under pure strategy**) *In the attack-defense game, the combination of strategy* (C_m, A_n) *with* $C_m \in S_1$ *and* $A_n \in S_2$ *is a Nash Equilibrium if and only if, for each player* k, *the action* C_m *or* A_n *is the best response to the other player:*

$$u_1(C_m, A_n) \geq u_1(C_i, A_n) \ for \ any \ C_i \in S_1$$

$$u_2(C_m, A_n) \geq u_2(C_m, A_j) \ for \ any \ A_j \in S_2$$

However, in the VANETs system, the pure strategies seldom happens since (C_m, A_n) means that the defender selects C_m as the only countermeasure and the attacker selects A_n as the only attack in the attack-defense game. Both sides of players wouldn't take this strategy in a long-term process in the VANETs security and privacy system. In a word, both defender and attacker will select actions with a certain probability distribution which compose a mixed strategy. The below is the definition of a mixed strategy.

Definition 2 *A **mixed strategy** for the attack-defense game is a strategy of selecting countermeasures with a probability distribution* $P_1 = (P_{C_1}, \cdots, P_{C_8})$, *where* $0 \leq P_{C_i}$ *and* $\sum_{i=1}^{8} P_{C_i} = 1$ *for defenders or* $P_2 = (P_{A_1}, \cdots, P_{A_8})$, *where* $0 \leq P_{A_j}$ *and* $\sum_{i=1}^{8} P_{A_i} = 1$ *for attackers. If player Def believes that player Att will play the strategies* S_2 *with probability* $P_2 = (P_{A_1}, \cdots, P_{A_8})$, *the expected payoff for player Def obtained with the pure strategy* C_i *is:*

$$\sum_{j=1}^{8} P_{A_j} ROI(C_i, A_j)$$

If player Att believes that player Def will play the strategies S_1 *with probability* $P_1 = (P_{C_1}, \cdots, P_{C_8})$, *the expected payoff for player Att obtained with the pure strategy* A_j *is:*

$$\sum_{i=1}^{8} P_{C_i} ROA(C_i, A_j).$$

Definition 3 (**Nash Equilibrium under mixed strategies**) *If the players Def and Att play respectively the strategies* S_{P_1} *with probability* $P_1 = (P_{C_1}, \cdots, P_{C_8})$, *and* S_{P_2} *with probability* $P_2 = (P_{A_1}, \cdots, P_{A_8})$, *the expected payoffs for the players are computed as follows:*

$$u_1(S_{P_1}, S_{P_2}) = \sum_{i=1}^{8} \sum_{j=1}^{8} P_{C_i} P_{A_j} ROI(C_i, A_j)$$

$$u_2(S_{P_1}, S_{P_2}) = \sum_{i=1}^{8} \sum_{j=1}^{8} P_{C_i} P_{A_j} ROA(C_i, A_j)$$

The mixed strategy $(S_{P_1^*}, S_{P_2^*})$ is a Nash Equilibrium only if the mixed strategy for each player is the best response to the mixed strategy of the other player:

$$u_1(S_{P_1^*}, S_{P_2^*}) \geq u_1(S_{P_1}, S_{P_2^*}) \, for \, any \, S_{P_1}$$

$$u_2(S_{P_1^*}, S_{P_2^*}) \geq u_2(S_{P_1^*}, S_{P_2}) \, for \, any \, S_{P_2}.$$

From the above two definitions, we can achieve the conditions of the Nash Equilibrium of mixed strategy $(S_{P_1^*}, S_{P_2^*})$:

$$\max \sum_{i=1}^{8} \sum_{j=1}^{8} P_{C_i} P_{A_j}^* ROI(C_i, A_j)$$
$$= \sum_{i=1}^{8} \sum_{j=1}^{8} P_{C_i}^* P_{A_j}^* ROI(C_i, A_j) \tag{4.1}$$

$$\max \sum_{i=1}^{8} \sum_{j=1}^{8} P_{C_i}^* P_{A_j} ROA(C_i, A_j)$$
$$= \sum_{i=1}^{8} \sum_{j=1}^{8} P_{C_i}^* P_{A_j}^* ROA(C_i, A_j) \tag{4.2}$$

where $P_{C_i}^* \in P_1^*, i = 1, 2, \cdots, 8; P_{A_j}^* \in P_2^*, j = 1, 2, \cdots, 8$.

However, it is a challenge to obtain the probabilities of the countermeasure and attack actions. This is because that we have to solve two groups of unknown probabilities by optimizing two payoff functions at the same time. To address this issue, based on the assumption that every participant is rational, we conclude the following Theorem 1, which provides a simplified solution to obtain the probabilities of countermeasure and the attack actions.

Theorem 1 *If $(S_{P_1^*}, S_{P_2^*})$ is the Nash Equilibrium of mixed strategies for the attack-defense game; p denotes the number of countermeasures $(C_{i_k}, k = 1, \ldots, p)$ taken by the defender with non-zero probabilities; q denotes the number of attacks $(A_{j_k}, k = 1, \ldots, q)$ launched by attacker with non-zero probabilities, then for any countermeasure C_{i_k} with probability $P_{C_{i_k}}$, the expected payoff of all attacks $(u_{A_{j_k}}, k = 1, \ldots, q)$ are equivalent for the attacker, vice versa.*

$$\sum_{k=1}^{p} P_{C_{i_k}} ROA(C_{i_k}, A_{j_1}) = \sum_{k=1}^{p} P_{C_{i_k}} ROA(C_{i_k}, A_{j_2})$$
$$= \cdots = \sum_{k=1}^{p} P_{C_{i_k}} ROA(C_{i_k}, A_{j_q}) \tag{4.3}$$

$$\sum_{k=1}^{q} P_{A_{j_k}} ROI(C_{i_1}, A_{j_k}) = \sum_{k=1}^{q} P_{A_{j_k}} ROI(C_{i_2}, A_{j_k})$$

$$= \cdots = \sum_{k=1}^{q} P_{A_{j_k}} ROI(C_{i_p}, A_{j_k}) \qquad (4.4)$$

Proof *Case I: Without loss of generality, we assume that the expected payoff $u_{A_{j_1}}$ = $\sum_{k=1}^{p} P_{C_{i_k}} ROA (C_{i_k}, A_{j_1})$ is less than the other expected payoffs in the attack-defense game. It indicates that with the probability of P_1^* the expected payoff of A_{j_1} is lower than that of other attack actions. Therefore, the attacker will not select A_{j_1} at all which leads $P_{A_{j_1}} = 0$. This is contradictive with the assumption of $P_{A_{j_1}}$ greater than 0.*

Case II: Without loss of generality, we assume that $u_{A_{j_1}}$ is greater than the other expected payoffs in the attack-defense game. It means that with the countermeasures' probabilities of P_1^* the expected payoff of A_{j_1} is higher than that of other attack actions. Therefore the attacker must select only A_{j_1} which leads $P_{A_j} = 1$. Under this condition, the defender must choose countermeasures with the max ROI related to A_{j_1} while set the probabilities of the other countermeasures as zero. This is contradictive with the assumption of p non-zero probabilities of countermeasures. ∎

We can achieve the Nash Equilibrium of mixed strategy from Theorem 1. However, the previous conclusion cannot reflect the impact of the cost and gain change of different actions on the chosen strategies of the participants (or Nash Equilibrium in the Attack-Defense Game). This is especially important for VANETs, which are typically a dynamic network with the frequently changed network architecture. Thus, a specific action may lead to a different cost as well as the gain under different environment setting. In the follows, we take the malicious node revocation as an example to show the change of costs and gains in different cases.

Case 1: A Low Defense Cost with A High Defense Gain In the case of presence of the network infrastructure (e.g., Road Side Unit), a node could easily revoke a malicious node by contacting the security authority via vehicle to RSU communications and the security authority could broadcast a revocation message within the whole network to revoke the target malicious node, which incurs a low defense cost and high defense gain.

Case 2: A Moderate Defense Cost with A Moderate Defense Gain In a case of of no infrastructure but the presence of sufficient number of legitimate users, the different vehicles could collaborate to revoke a malicious node, which leads a moderate defense cost (e.g. transmission of coordinate messages among the different collaborative nodes) as well as a moderate defense gain (e.g., local revocation of this malicious node rather than global revocation) with a certain successful rate (e.g., revocation failure if no enough voting numbers).

Case 3: A High Defense Cost with A Low Defense Gain In a case of even no enough collaboration nodes, a legitimate user can still revoke a node by launching a suicide revocation, which incurs a high defense cost (e.g., also revoking its own public/private key) with a low defense gain (e.g., a limited number of transmission range of revocation message in a sparse network).

From the above example, it is obvious that the cost and gain of the defender could have a significant change in different scenarios, which has a direct impact on the strategy choosing of the attacker and the defense. In the follows, we use Theorem 2 to model the impact of the change of the gains and the costs on the Nash Equilibrium.

Theorem 2 *If the gain or the cost incurred by a specific attack takes change and this change leads to an increased utility of the attacker (e.g., a higher R_A or lower RM), the defender will perform the countermeasures corresponding to this attack with a higher probability. Conversely, if the gain or the cost incurred by a specific countermeasure takes change and this change leads to an increased utility of the defender (e.g., a higher R_D or higher RM), the attacker will perform the attacks regarding to this countermeasure with a lower probability.*

Proof *Without loss of generality, we suppose for a specific attack A_{j_1}, $C_{i_k}(k = 1, \cdots, r)$ are the countermeasures which can mitigate A_{j_1}; $C_{i_k}(k = r+1, \cdots, p)$ are the countermeasures which cannot mitigate A_{j_1}. For $C_{i_k}(k = r+1, \cdots, p)$, their $ROA(C_{i_k}, A_{j_1})$s are equivalent, denoted by $ROA(A_{j_1})$ and we conclude that $ROA(A_{j_1})$ is greater than any $ROA(C_{i_k}, A_{j_1})(k = 1, \cdots, r)$ from the Eq. (4.3). The expected payoff of A_{j_1} is:*

$$u_{A_{j_1}} = \sum_{k=1}^{r} P_{C_{i_k}} ROA(C_{i_k}, A_{j_1}) + \sum_{k=r+1}^{p} P_{C_{si_k}} ROA(A_{j_1})$$

$$= ROA(A_{j_1}) + \sum_{k=1}^{r} P_{C_{i_k}} [ROA(C_{i_k}, A_{j_1}) - ROA(A_{j_1})]$$

If $R_{A_{j_1}}$ is increased(or RM is decreased), consequently $u_{A_{j_1}}$ will be increased. To keep the Eq. (4.3) in Theorem 1, the defender has to increase $P_{C_{i_k}}(k = 1, \cdots, r)$.

We also suppose for a specific countermeasure C_{i_1}, $A_{j_k}(k = 1, \cdots, r)$ are the attacks which can be mitigated by C_{i_1}; $A_{j_k}(k = r+1, \cdots, q)$ are the attacks which cannot be mitigated by C_{i_1}. For $A_{j_k}(k = r+1, \cdots, q)$, their $ROI(C_{i_1}, A_{j_k})$s are equivalent to -1, and we conclude that -1 is great than any $ROI(C_{i_1}, A_{j_k})(k = 1, \cdots, r)$ from the eq. (4.1). The expected payoff of C_{i_1} is:

$$u_{C_{i_1}} = \sum_{k=1}^{r} P_{A_{j_k}} ROI(C_{i_1}, A_{j_k}) + \sum_{k=r+1}^{q} P_{A_{j_k}}(-1)$$

$$= \sum_{k=1}^{r} P_{A_{j_k}} [ROI(C_{i_1}, A_{j_k}) + 1] - 1$$

If $R_{D_{i_1}}$ is increased(or RM is increased), consequently $u_{C_{i_1}}$ will be increased. To keep the Eq. (4.4) in Theorem 1, the attacker has to decrease $P_{A_{j_k}}(k = 1, \cdots, r)$. ∎

Theorem 2 shows that in the attack-defense game, if the attacker can get more payoffs from an attack or use this attack more easily than other, the defender must choose the countermeasures related to this attack with a higher priority to avoid the failure of the defense; if the defender can effectively mitigate an attack, the attacker must decrease the possibility of launching this attack, or even give up using this attack.

4.3 Security Analysis of Attack-Defense Game: A Case Study

In this section, we investigate an attack-defense game to illustrate the Theorem 1 and Theorem 2 by using the specific attack-defense tree for VANETs security and privacy presented in Chap. 3. For this specific case study, the detailed payoff values of attacker and defender are given in Table 4.2. In particular, Tables 3.1 and 3.2 in Chap. 3 summarize all the factors to calculate the payoff of defender (ROI) and payoff of attacker (ROA), respectively.

Before we solve the Nash Equilibrium of this Attack-defence game, it is useful to introduce the concept of *Dominated Strategy* to simplify the process of solving final solutions. A Dominated strategy means that its'payoff is less than any other strategy's payoffs under the same strategy of the adversary in the attack-defense game.

From Table 4.2, we observe that the attacks of AS_2 or AS_3 are dominated strategies for attackers. The countermeasures of C_2, C_4, C_5, C_6 and C_7 are dominated strategies for defenders. The reduction solutions are shown in Table 4.3. This is a mixed strategy for both defender and attacker. The probabilities of choosing attacks of AS_2 or AS_3 are zeros for attackers. Similarly probabilities of choosing countermeasures of C_2, C_4, C_5, C_6 and C_7 are zeros for defenders. From Theorem 1, for any countermeasure C_i with probability P_{Ci}, the expected payoff of the attacker in any attack are equivalent, vice versa, we therefore get the equations as follows:

$$(-1)P_{C_1} = 0.5(P_{C_1} + P_{C_8}) + (-0.875)P_{C_3}$$
$$= 0.333(P_{C_1} + P_{C_3}) + (-1)P_{C_8}$$

(4.5)

$$P_{AS_1} + (-1)(P_{AP_{11}} + P_{AP_{12}} + P_{AP_{13}} + P_{AP_2} + P_{AP_3})$$
$$= (-1)(P_{AS_1} + P_{AP_2} + P_{AP_3})$$
$$+ (-0.625)(P_{AP_{11}} + P_{AP_{12}} + P_{AP_{13}})$$
$$= (-1)(P_{AS_1} + P_{AP_{11}} + P_{AP_{12}} + P_{AP_{13}} + P_{AP_3})$$

(4.6)

From above equations, we can obtain the two groups of probabilities of mixed strategies for the attacker and defender: $P_{AS_1} = 3/25$, $P_{AP_1} = 16/25$, $P_{AP_2} = 6/25$, $P_{AP_3} = 0$; and $P_{C_1} = 17/109$, $P_{C_3} = 52/109$, $P_{C_8} = 40/109$. Here, $P_{AP_3} = 0$ indicates that the attack AP_3 is also a dominated strategy for attackers. Until now we have obtained the Nash Equilibrium of this attack-defense game. That is for the attacker taking the attacks of AS_1, AP_1 and AP_2 with the probabilities of 3/25, 16/25 and 3/25 respectively; for the defender choosing the countermeasures of C_1, C_3, and C_8 with the probabilities of 17/109, 52/109 and 40/109 respectively.

According to the final Equilibrium results, we find that the defender will adopt the countermeasure C_3 with the highest probability of 52/109; the attacker will choose the attack AP_1 with the highest probability of 16/25. These solutions indicate that in our considered VANETs security scenario, the defender will cooperate with the other vehicles (countermeasure C_3) as far as possible to mitigate the attacks of tracing identity (AP_1, including AP_{11}, AP_{12} and AP_{13}), and vice versa. In fact, due

Table 4.2 The utilities of participants in attack-defense game

	AS_1	AS_2	AS_3	AP_{11}	AP_{12}	AP_{13}	AP_2	AP_3
C_1	1,−1	−1,−0.6	−1,−0.667	−1,0.5	−1,0.5	−1,0.5	−1,0.333	−1,−0.2
C_2	−1,0	−0.375,−0.786	−1,−0.667	−1,0.5	−1,0.5	−1,0.5	−1,0.333	−1,−0.2
C_3	−1,0	−0.375,−0.889	−1,−0.667	−0.625,−0.875	−0.625,−0.875	−0.625,−0.875	−1,0.333	−1,−0.2
C_4	−1,0	−0.375,−0.786	−1,−0.667	−1,0.5	−1,0.5	−1,0.5	−1,0.333	−1,−0.2
C_5	−1,0	−1,−0.6	−1,−0.667	−0.75,−0.55	−1,0.5	−1,0.5	−1,0.333	−1,−0.2
C_6	−1,0	−1,−0.6	−1,−0.667	−1,0.5	−0.75,−0.55	−1,0.5	−1,0.333	−1,−0.2
C_7	−1,0	−1,−0.6	−1,−0.667	−1,0.5	−1,0.5	−0.75,−0.55	−1,0.333	−1,−0.2
C_8	−1,0	−1,−0.6	−1,−0.667	−1,0.5	−1,0.5	−1,0.5	0,−1	−1,−0.2

Table 4.3 Reduction of attack-defense game

	AS_1	AP_1	AP_2	AP_3
C_1	1,−1	−1,0.5	−1,0.333	−1,−0.2
C_3	−1,0	−0.625,−0.875	−1,0.333	−1,−0.2
C_8	−1,0	−1,0.5	0,−1	−1,−0.2

to the high costs of stealing identity to utilize insider false data (AS_2) and using the camera to impinge privacy (AP_3), the attacker has to give up these two attacks in this attack-defense game (the probabilities of AS_2 and AP_3 are zeroes in the Nash Equilibrium).

Figure 4.1 illustrates how the attack gain-cost ratio (R_A)and effectiveness of defense strategies (RM) affect the result of Nash Equilibrium of mixed strategies. In case 1: $R_A = 0$ indicates that the attacker won't have any gain (or failure attack) by launching the specific attack AS_1. For example, a attacker may be malicious who seeks no personal benefits from the attacks but aims to disrupt the VANETs security. Since in our game, the utilities of both sides are known by each participant, the defender chooses a low probability ($P_{C_1} = 0.15$) of countermeasure C_1. Under this circumstance, the probabilities of countermeasures C_1, C_3 and C_8 keep no changing whatever RM is low or high. This is reasonable since this attack won't affect the security of VANETs system with $R_A = 0$.

In case 2: $R_A = 1$ means that the attacker increased his gain by launching the attack AS_1 comparing with that in case 1. Under this condition, the probability of countermeasure C_1 although is decreasing with the more effective countermeasure C_1 (RM increasing), it is increased ($P_{C_1} > 0.15$) comparing with that in case 1 when the attacker have no gain($P_{C_1} = 0.15$). This has shown the result of Theorem 2: a higher gain of attacker, a higher probability of corresponding countermeasure in Nash Equilibrium of mixed strategies.

In case 3: $R_A = 10$ means a high gain-cost ratio for the attacker. Under this condition the defender will choose countermeasure C_1 with probability of 1 as long as RM is less than 0.85. However as RM tends to 1 (i.e. countermeasure C_1 is 100 % effective) the probability of C_1 decreases to 0.15. Due to the relationship of $P_{C_1} + P_{C_3} + P_{C_8} = 1$, the probabilities of countermeasures of C_3 and C_8 are zeros when RM is less than 0.85. Both of them increase when RM is greater than 0.85. This result explains that when the countermeasure C_1 is very effective(RM close to 1), the attacker will choose other attacks AP_1 or AP_2 with higher probabilities which lead the defender to choosing the corresponding countermeasures C_3 and C_8 with higher probabilities.

Figure 4.2 illustrates the impacts on probabilities of attacks brought by changes of defender's gain-coast ration R_D and effectiveness of countermeasures RM. In case 1: $R_D = 0$ means that the defender have no gain (or failure defense) by using countermeasure C_1. The attacker will launch the attack AS_1 of probability 1 since defender's low utility will lead a low probability of countermeasure C_1.

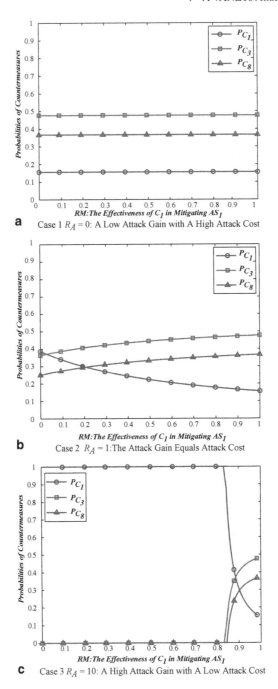

Fig. 4.1 Probabilities of Countermeasures with the changes of R_A and RM of Nash Equilibrium: The lines with circles, squares and triangles represent the probabilities of P_{C_1}, P_{C_3} and P_{C_8}, respectively. **a** Case 1 $R_A = 0$: A Low Attack Gain with A High Attack Cost, **b** Case 2 $R_A = 1$: The Attack Gain Equals Attack Cost, **c** Case 3 $R_A = 10$: A High Attack Gain with A Low Attack Cost

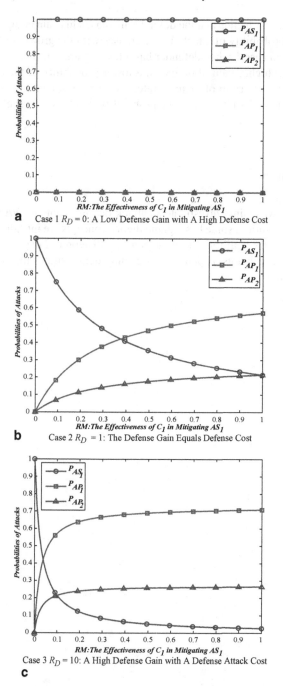

Fig. 4.2 Probabilities of Attacks with the changes of R_D and RM of Nash Equilibrium: The lines with circles, squares and triangles represent the probabilities of P_{AS_1}, P_{AP_1} and P_{AP_2}, respectively. **a** Case 1 $R_D = 0$: A Low Defense Gain with A High Defense Cost, **b** Case 2 $R_D = 1$: The Defense Gain Equals Defense Cost, **c** Case 3 $R_D = 10$: A High Defense Gain with A Defense Attack Cost

In case 2: $R_D = 1$ means that the defender increased his gain comparing with that in case 1. The probability of attack AP_1 decreases with the growing of RM and it is lower than that in case 1 when defender has a lower gain. This has shown the result of Theorem 2: a higher gain of defender, a lower probability of the corresponding attack in Nash Equilibrium of mixed strategies. In case 3, under the condition of a higher gain of defender ($R_D = 10$), the probability of attack AP_1 is even lower than that in case 2.

4.4   Summary

In this Chapter, we mainly investigate an attack-defense game for VANETs security and privacy. In addition, some basic game theory concepts are introduced as needed. Theorem 1 shows the equilibriums of the game and Theorem 2 explains the strategies using trend of participants. A detailed case study demonstrates the effectiveness of the game solutions.

Chapter 5
Modelling of Multiple Phased Attack on VANET Security

Abstract Many real world systems operate in phased mission where the reliability structure varies over consecutive time periods, known as phases. For mission success, all phases must be completed without failure. In this Chapter, we use a phased attack-defense tree to model the actions of the attacker and the defender for VANETs communication security. Based on the phased attack-defense tree model, we further investigate a phased attack-defense game which interprets the interactions and dependency between the attacker and defender. A brief solution of the phased game is presented as an illustration for the sequence actions of attacker and defender under certain situations.

5.1 System Architecture

In this study, we consider a VANET attack-defense process as a phased mission system due to the following reasons. Firstly, the attack strategy adopted by the attackers highly depends on the defense strategies of the defender. Intuitively, if there is no defense system at all, the attacker will prefer to launching a low-cost but high-revenue attack rather than performing a high-cost but low-revenue attack. Therefore, in practice, the attack strategy will improve from simple attack to more advanced yet complicated attacks along with the evolution of the defense strategy. This process could be naturally modelled as a phased mission system. In the first phase in which no security defense is considered, the attacks start from the low-cost and simple attacks. In the subsequent phases, along with the enhanced defense strategy, the attacks become more and more complicated with an increased cost. Therefore, the defending work will succeed if and only if all of phases succeed. In the following, we will illustrate the phased mission system based on attack-defense tree for security problems in VANETs.

In VANETs system, we build a model called the *phased attack-defense tree* for the security assessment. The security in communication is considered as the target in this study. We suppose the situation in which the attacker broadcasts bogus information in the network to mislead or even affect the behavior of other drivers. For example, both normal users and the attacker are driving to a destination through two ways of different distances. The attacker can broadcast the bogus information such as "the nearer road is jammed", then the normal users who believe it will choose the longer road. That means the attacker can mislead normal users to choose the longer way thus free the nearer way for himself. Conversely, the normal users deceived by the bogus information will suffer more distance. We set the *Compromise Communication Security* (denoted as T) as the top goal of the phased attack-defense tree.

In this study, we use three phases to demonstrate the progressive actions of attack-defense process. In each phase, each pair of actions includes only a certain attack action and its countermeasure corresponding with each other. When the attack and defense proceed from the previous phase to the next, the defense countermeasures in a previous phase are the destroyed targets of attacks in the next phase. The attacker must select more advanced attacks in the next phase directly against the existing countermeasures. Also, in each phase, the attacker has his own sub-goals for getting the final objective.

Three sub-goals in each phase of T are *Bogus Information, Insider False Data Injection* and *Cooperative Attack*, which are denoted as B, I and C, respectively. In other words, an attacker could achieve the attack objective T by compromising any phase of the compromised communication security of normal users in each sub-goal. The phased attack-defense tree for VANETs security is shown in Fig. 5.1.

In *Security Phase 1*, to launch the bogus information, the attacker must take two actions: misleading information and broadcasting. The countermeasure corresponding to phase 1 attack is *PKI & Signature*, denoted as P, which means that by requiring each vehicle to provide the public key and signature for each sending message, the security authority could prevent the bogus information attack from broadcasting unauthorized messages.

In *Security Phase 2*, to launch the insider injection false data, it is necessary for the attacker to perform two attack steps of compromising key and bogus information, in which an attacker could compromise a legitimate node's primary key and disseminate the unauthorised information by using the compromised key. To thwart fake attack, the security authority could use the countermeasure called *voting* (V) which means that the defender could crosscheck the information under the collaboration of multiple nodes, and send the reports to the Certificate Authority to revoke the public key of a misbehaving vehicle.

In *Security Phase 3*, to launch the cooperative attack, the attacker still adopts insider false data injection by cooperating with other misbehaving users. The attacker and his partners could play a leading role in the crosscheck to make the defenders believe in the misleading message. When the defender finds the loss brought by the false data, he will send a request to the Certificate Authority for revoking the public keys of himself and the attacker, synchronously. We call this countermeasure *Suicide*, which is the last method for preserving the security from the strongest attack.

From the description of phased attack-defense tree, we can see that it has three phases, which is different from the previous attack-defense tree model. The former tree model only considers one phase in which the attacker and defender act all potential attacks and countermeasures simultaneously. Hence, we will introduce a

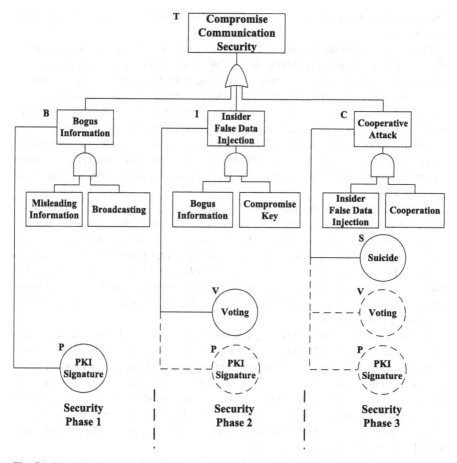

Fig. 5.1 Phased attack-defense tree

phased attack-defense game model to investigate the strategies of participants based on the phased attack-defense tree in Sect. 5.2.

5.2   A Phased Attack-Defense Game for VANETs

In the previous section, we build the phased attack-defense tree to model the security and privacy issues in VANETs. We assume the attacker and defender as rational nodes when considering their utilities.

In Chap. 4, we describe the attack-defense game as a static game of complete information, because we suppose that each participant would take action simultaneously with common knowledge. Differently, in this chapter we will present a new game called *the phased attack-defense game* based on the phased attack-defense tree.

The phased attack-and-defense game model differs from the previous static game in two aspects.

One aspect is that the phased attack-defense game is a *dynamic game* with three phases. In Phase 1, when launching *Bogus Information*, the attacker does not know whether the defender adopts a countermeasure of *PKI & Signature*; and vice versa. Both of their actions can be considered as occurring simultaneously because they only perceive the rival's behaviors at the end of the game. But, in the third phase, based on the phased attack-and-defense tree, the defender can observe the attacker's moves before his action. Thus, in Phase 3, the attacker's and the defender's moves occur in sequence, because after observing the attacker's *Cooperative Attack*, the defender can choose the countermeasure *Suicide*. This indicates that the phased game is a *dynamic game*.

Another aspect is that the phased attack-defense game is of *imperfect information*, though the attacker can know the full history of the actions of the game, the defender cannot know the history of the attack's moves. A previous attack which did not succeed sometimes cannot be observed by the defender before the next attack in our case. For example, while the attacker launch *Insider False Data Injection* in the Phase 2, the defender cannot know whether *Bogus Information* in Phase 1 had been acted or not, unless it has succeeded. So in this chapter, the presented game will be of *imperfect information*.

As the same as the previous attack-defense game, the participants' payoffs are always common knowledge thus this game is of *complete information* . That means, each player's payoff function(the function that determines the player's payoff from the combination of actions chosen by the players) is common knowledge among all the players. For example, we suppose the attacker and defender know the payoff and the loss in a traffic jam. Because they can set the saved or wasted time as their utilities, the average driving time lengths of two different roads are the common knowledge among attacker and defender.

In summary, the phased attack-defense game model is a three-phase dynamic game of complete but imperfect information.

5.2.1 Game Model

The game model can be used to analyze the probabilities of their strategies according to the utility functions. We can get the equilibriums of the optimal results by using backward induction method. We describe this game as follows. The game $\mathcal{G}$ can be defined as a triplet $(\mathcal{O}; \mathcal{S}; \mathcal{U})$, where $\mathcal{O}$ is a set of players, $\mathcal{S}$ is a set of strategies and $\mathcal{U}$ is a set of payoff functions.

- **Players**: The set of $\mathcal{O} = \{O_i\}$ includes an attacker O_1 and a defender O_2, referring to *Att and Def respectively.*
- **Strategy**: The set of $\mathcal{S} = \{S_i, i = 1, 2\}$ includes the strategies of the participants in three different phases, in which S_i contains the behaviors with the probabilities.

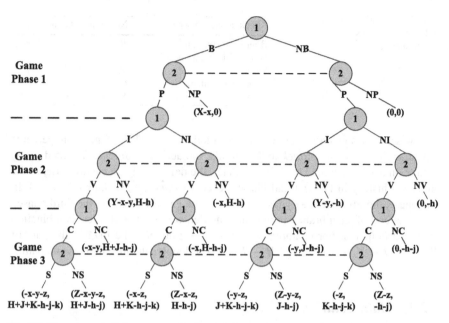

Fig. 5.2 Phased attack-defense game: B, I and C denote the actions of attacker, and P, V and C denote the actions of defender; a combination e.g., $(Y - x - y, H - h)$ denotes the final utilities of attacker and defender, respectively.

- **Payoff function**: The set of $\mathcal{U} = \{U_i, i = 1, 2\}$ of the utilities consists of U_1, which is the sum of the return of attacks (*ROA*) by attacker and U_2 which is the sum of the return of investment (*ROI*) by defender in every complete attack-defense process.

Figure 5.2 shows the phased attack-defense game model. The notations, gains and costs of actions are shown in Table 5.1.

In *Game Phase 1*, the attacker can launch the attack *Bogus Information* (denoted by B) or give it up (denoted by NB), because he may quit this game or launch an attack in next phase. Simultaneously, the defender would consider adopting the countermeasure *PKI & Signature* (denoted by P) or give it up (denoted by NP). In this phase, we can see that it is a static process. If the attacker could succeed by the attack or quit at first, this game will end; else, the game will enter the second phase.

In *Game Phase 2*, based on the countermeasure *PKI & Signature* of defender, the attacker can choose the attack *Insider False Data Injection* (denoted by I) or give it up (denoted by NI). Meanwhile, the defender would determine to adopt the *Voting* (denoted by V) mechanism or not to do (denoted by NV). In this phase, it is also a static situation, but differently, the defender cannot observe the history of the attacker in the first phase. The information is imperfect to the defender.

In *Game Phase 3*, exploiting the *Voting* mechanism, the attacker can collaborate with other malicious users to launch the attack *Cooperative Attack* (denoted by C),

Table 5.1 Notations, gains
and costs of attacks and
countermeasures

Attack/countermeasure	Notation	Gain	Cost
Bogus information	B	X	x
Insider false data injection	I	Y	y
Voting	V	J	j
Cooperative attack	C	Z	z
Suicide	S	K	k

otherwise, he has no choice but to quit (denoted by NC). The defender can perceive this attack *Cooperative Attack* and then take the action of *Suicide* (denoted by S) or take on action (denoted by NS). Similarly, the defender again cannot observe the attacker's history in Phase 1 and Phase 2 except *Cooperative Attack* in Phase 3. In this phase, therefore, it is a dynamic process. The game will end in the third phase.

At the end of each branch in the game model, we give the utilities' combination of the participants. For example, $Z - x - y - z$ denotes the *ROA* in the path of $B - P - I - V - C-NS$, and $H + J - h - j$ denotes the *ROI* in the same path.

5.2.2 Main Solutions

In this study, the varying gains and costs of both participants will bring about the diverse results of equilibrium of this game. We will illustrate a brief solution under certain condition of comparisons of gains and costs in each phase. With the similar analysis, we can get other equilibriums under the all possible comparisons of gains and costs of all actions. In Phase 3, we assume that the gains of both attacker and defender are lower than their costs, (i.e. $Z < z$, and $K < k$). This assumption is reasonable because for defender choosing *Suicide* is a high cost action; and for attacker, it is difficult to find more cooperators to launch *Cooperative Attack*. In both Phase 1 and Phase 2, we assume that the gain of each action is higher than the cost, (i.e. $X > x$, $H > h$, $Y > y$, $J > j$) because the attacker can easily launch *Bogus Information* and *Insider False Data Injection* and the countermeasures *PKI & Signature* and *Voting* are also worth acting for the defender. In summary, we assume as follows:

$$X > x, H > h, Y > y, J > j, Z < z, K < k. \tag{5.1}$$

Because our phased attack-defense game is a dynamic game of complete but imperfect information, the defender cannot observe the actions of attacker in previous phases. In other words, the defender in Phase 3, needs to estimate the possibilities of actions of the attacker including *Bogus Information*, and *Insider False Data Injection*. For solving the equilibrium of the phased attack-defense game, we use the backward induction method. According to $K < k$ or $K - k < 0$ of the assumption (5.1), we can get $H + J + K - h - j - k < H + J - h - j$, which means that for the defender, the utility of giving up *Suicide* is higher than that of choosing this countermeasure. Therefore, the defender will choose giving up the countermeasure with the same

Table 5.2 The utility matrix of phase 2 in phased attack-defense game

			Defender	
			V	NV
Attacker	Left	I	$-x-y, H+J-h-j$	$Y-x-y, H-h$
		NI	$-x, H-h-j$	$-x, H-h$
	Right	I	$-y, J-h-j$	$Y-y, -h$
		NI	$0, -h-j$	$0, -h$

branches to avoid more loss of its utilities. By knowing the defender's strategy of *NS*, the attacker will not choose launching *Cooperative Attack* in all branches, for the same reason that the gain is lower than the cost (e.g. in the leftmost selecting set, $Z-x-y-z < -x-y$). In a word, both the attacker and the defender maximize their utilities in backward sequence. The solution of Phase 3 is that the attacker chooses the branch of *NC*, and the defender does not need to *suicide*.

We now consider the actions of Phase 2. In Phase 2 the attacker and the defender choose the actions, simultaneously. There are two different situations. One is that the attacker chooses *Bogus Information*; the other is that the attacker gives up *Bogus Information*. The description of utilities in Phase 2 is shown in Table 5.2. We can give the part of solutions by using the mixed strategies equilibrium in the left as follows:

$$p_I(H + J - h - j) + (1 - p_B)(H - h - j) = H - h$$

$$p_V(-x - y) + (1 - p_V)(Y - x - y) = -x$$

By solving the two equations, we get the probability of *Insider False Data Injection*: $p_I = j/J$, and the probability of *Voting*: $p_V = 1 - y/Y$. The combinations of utilities are $(-x, H - h)$ in the left branches and $(0, -h)$ in the right branches in Phase 2, respectively.

The results indicate that the higher cost j (or the lower gain J) of the countermeasure *Voting*, the higher probability of *Insider False Data Injection* the attacker will choose. This solution is reasonable, since when the defender has more chance to give up the high cost and low gain defense countermeasure *Voting*, the attacker will prefer to launch an attack *Insider False Data Injection*. On the contrary, the higher cost y (or the lower gain Y) of the attack *Insider False Data Injection*, the lower probability of *Voting* the defender will choose. By considering the attacker's choice of giving up the high cost and low gain attack *Insider False Data Injection*, the defender will have a lower probability of using a countermeasure *Voting*.

We get the expected utilities of attacker and defender in Phase 2, which also are the expected utilities of the paths $B - P$ and $NB-P$, respectively. So we can give the description of utilities in Phase 1 shown in Table 5.3. Similarly, by solving the mixed strategies equilibrium, the solution is that the probability of *Bogus Information*: $p_B = h/H$ and the probability of *PKI & Signature*: $p_P = 1 - x/X$. Readers can get the similar explanations of this result as that in Phase 2.

In summary, in Phase 1, the attacker chooses *Bogus Information* with the probability of h/H, and the defender chooses *PKI & Signature* with the probability of $1-x/X$; in Phase 2, the attacker chooses *Inside False Data Injection* with the probability of

Table 5.3 The Utility Matrix
of Phase 1 in Phased
Attack-Defense Game

		Defender	
		P	*NP*
Attacker	*B*	$-x, H-h$	$X-x, 0$
	NB	$0, -h$	$0, 0$

j/J, and the defender chooses *Voting* with the probability of $1-y/Y$; in Phase 3, the attacker gives up *Cooperative Attack*, and the defender gives up *Suicide*. We can find that in Phase 1 and Phase 2, the strategies of participants depend on the rival's gains and costs, but in Phase 3, the strategies only depend on their own gains and costs.

5.3 Summary

In this Chapter, we use a phased attack-defense tree model to describe the behaviors of attacker and defender in VANETs communication security problem. A phased dynamic game is investigated further to show the dependency between attacker and defenders strategies. Finally, a brief solution of the dynamic game is given to illustrate the sequence actions of participants under certain cases.

Glossary

Attack-Defense Game	An attack-defense game is a game where the players are an attacker and a defender for security in a system.
Attack-Defense Tree	An attack-defense tree is a conceptual diagram showing how an asset, or target, might be attacked and how the defense can be acted.
Attack Tree	An attack tree is a conceptual diagram showing how an asset, or target, might be attacked.
Authentication	Authentication is the act of confirming the truth of an attribute of a datum or entity. This might involve confirming the identity of a person or software program, tracing the origins of an artifact, or ensuring that a product is what its packaging and labeling claims to be.
Content Distribution	Content distribution is a approach of proxy-servers located at strategic points around the Internet and arranged so as to ensure that a download request can always be handled from the nearest server.
Denial-of-Service	Denial-of-service (DoS) is an attempt to make a machine or network resource unavailable to its intended users.
Dynamic Game	A dynamic game is a game where one player chooses his action before the others choose theirs.
Elliptic Curve Digital Signature Algorithm	Elliptic curve digital signature algorithm (ECDSA) is a variant of the Digital Signature Algorithm (DSA) which uses elliptic curve cryptography.
Game of Complete Information	A game of complete information is a game where each player's payoff function is common knowledge among all the players.
Game of Imperfect Information	A game of imperfect information is a game where one player cannot know the history of the other's moves.

S. Du, H. Zhu, *Security Assessment in Vehicular Networks,* SpringerBriefs in Computer Science, DOI 10.1007/978-1-4614-9357-0, © The Author(s) 2013

Location-based Service	Location-based service (LBS) is a general class of computer program-level services used to include specific controls for location and time data as control features in computer programs.
Mixed Strategy	A mixed strategy is an assignment of a probability to each pure strategy. This allows for a player to randomly select a pure strategy.
Nash Equilibrium	A set of strategies is a Nash equilibrium if no player can do better by unilaterally changing his or her strategy.
Payoff Function	The function determines the player's payoff from the combination of actions chosen by the players.
Phased Attack-Defense Game	An phased attack-defense game is a attack-defense game where the attack-defense process has not only one phase.
Public Key Infrastructure	A public key infrastructure (PKI) is an arrangement that binds public keys with respective user identities by means of a certificate authority (CA).
Pure Strategy	A pure strategy provides a complete definition of how a player will play a game. In particular, it determines the move a player will make for any situation he or she could face.
Return on Attack	Return on attack is the ratio of money gained or lost (whether realized or unrealized) on an attack relative to the amount of money costed.
Return on Investment	Return on investment is the ratio of money gained or lost (whether realized or unrealized) on an investment relative to the amount of money invested.
Risk Mitigation	Risk mitigation is the capacity of reducing the success rate of attack.
Static Game	A static game is a game where the players simultaneously choose actions and the players receive payoffs that depend on the combination of actions just chosen.

References

1. Watson, Jason D., et al, "Simulation and Analysis of Extended Brake Lights for Inter-Vehicle Communication Networks", *Distributed Computing Systems Workshops, 2007. ICDCSW'07. 27th International Conference on*. IEEE, 2007.
2. X. Lin, R. Lu, X. Liang and X. Shen, "STAP: A Social-Tier-Assisted Packet Forwarding Protocol for Achieving Receiver-Location Privacy Preservation in VANETs," Proc. *IEEE INFOCOM'*11, Shanghai, China, April 10–15, 2011.
3. X. Lin, R. Lu, C. Zhang, H. Zhu, P.-H. Ho and X. Shen, "Security in vehicular Ad Hoc networks," *IEEE Communications Magazine*, Vol. 46, No. 4, pp. 88–95, 2008.
4. Arbabi, Mohammad Hadi, and Michele Weigle. "Using vehicular networks to collect common traffic data", Proceedings of *the sixth ACM international workshop on VehiculAr InterNETworking. ACM*, 2009.
5. H. Zhu, R. Lu, X. Lin and X. Shen, "Security in Service-Oriented Vehicular Networks," *IEEE Wireless Communication Magazine*, vol. 16, no. 4, pp. 16–22, August, 2009.
6. Spyropoulos T, Psounis K, Raghavendra C S, "Efficient routing in intermittently connected mobile networks: the multiple-copy case", *Networking, IEEE/ACM Transactions on* 16.1, 77–90, 2008.
7. H. Hsiao, A. Studer, C. Chen, A. Perrig, F. Bai, B. Bellur, A. Lyer, "Flooding-resilient broadcast authentication for vanets", Proc. *ACM MOBICOM'11*, 2011.
8. Q. Han, S. Du, D. Ren and H. Zhu, "SAS: A Secure Data Aggregation Scheme in Vehicular Sensing Networks", *International Conference on Communications (IEEE ICC'10)*, Cape Town, South Africa, May 23–27, 2010.
9. S. Reidt, M. Srivatsa, S. Balfe, "The Fable of the Bees: Incentivizing Robust Revocation Decision Making in Ad Hoc Networks," Proc. *ACM CCS'09*, 2009.
10. M. RAYA, and J.-P.HUBAUX, "Securing vehicular ad hoc networks," JCS-SASN, 2007.
11. K. Hoeper and G. Gong, "Bootstrapping Security in Mobile Ad Hoc Networks Using Identity-Based Schemes with Key Revocation," Technical Report CACR 2006–04, Centre for Applied Cryptographic Research (CACR) at the University of Waterloo, Canada, 2006.
12. B.J. Matt. "Toward Hierarchical Identity-based Cryptography for Tactical Networks," In Proceedings of the *2004 Military Communications Conference (MILCOM 2003)*, pages 727–735. IEEE Computer Society, November 2004.
13. Y. Zhang, W. Liu, W. Lou, Y. Fang, and Y. Kwon, "AC-PKI: Anonymous and Certificateless Public Key Infrastructure for Mobile Ad Hoc Networks," In Proceedings of the *International Conference on Communications (ICC 2005)*, pages 3515–3519. IEEE Computer Society, May 2005.
14. J. Luo, J.-P. Hubaux, and P.T. Eugster, "DICTATE: DIstributed CerTification Authority with probabilisTic frEshness for Ad Hoc Networks," *IEEE Transactions on Dependable and Secure Computing*, 2(4):311–323, 2005.
15. J. Clulow and T. Moore, "Suicide for the Common Good: A New Strategy for Credential Revocation in Self-organizing Systems," ACM *SIGOPS Operating Systems Reviews*,40(3):18–21, 2006.

16. R. Anderson T. Moore, J. Clulow and S. Nagaraja, "New Strategies for Revocation in Ad-Hoc Networks," In Proceedings of the 4th *European Workshop on Security and Privacy in Ad Hoc and Sensor Networks (ESAS 2007)*, pages 232–246. Springer, July 2007.

17. T. Moore, M. Raya, J. Clulow, P. Papadimitratos, R. Anderson, and J-P. Hubaux, "Fast Exclusion of Errant Devices From Vehicular Networks," In Proceedings of the *5th conference on Sensor, Mesh and Ad Hoc Communications and Networks (SECON 2008)*, pages 135–143, 2008.

18. M. Raya, M. Hossein Manshaei, M. Fiäelegyhazi, and J-P. Hubaux, "Revocation Games In Ephemeral Networks," In Proceedings of the 15th *ACM conference on Computer and Communications Security*, pages 199–210. ACM, 2008.

19. J. Freudiger, M. Manshaei, J.-P. Hubaux, and D.C. Parkes, "On non-cooperative location privacy: A Game-Theoretic Analysis," *CCS'09*, 2009.

20. A. R. Beresford and F. Stajano, "Location privacy in pervasive computing. Pervasive Computing," *IEEE*, vol. 2, no. 1, 46–55, 2003.

21. S. Bistarelli, M. Dall' Aglio and P. Peretti, "Strategic games on defense trees," *Formal Aspects in Security and Trust*, Vol. 4691, 1–15, 2007.

22. D. Ren, S. Du and H. Zhu, "A Novel Attack Tree Based Risk Assessment Approach for Location Privacy Preservation in the VANETs," in Proc. of *ICC 2011*, 2011.

23. B. Kordy, S. Mauw, M. Melissen, and P. Schweitzer, "Attack–Defense Trees and Two-Player Binary Zero-Sum Extensive Form Games Are Equivalent," *DECISION AND GAME THEORY FOR SECURITY*, Vol. 6442, 245–256, 2010.

24. C. Cuyu, X. Yong, S. Meilin, "Development and status of vehicular ad hoc networks", *Journal on Communications*, no. 28, vol. 17, pp. 116–126, 2007.

25. J. Krumm, "Inference attacks on location tracks", in Proc. of*the 5th International Conference on Pervasion Computing*, Toronto, Ontario, Canada. Springer-Verlag, 2007, 127–143.

26. C. Troncoso, G. Danezis, "The Bayesian traffic analysis of mix networks", in Proc. of *the 16th ACM Conference on Computer and Community Security*. ACM CCS09, 369–379, 2009.

27. T. Leinmuller, E. Schoch and C. Maihofer, "Security requirements and solution concepts in vehicular ad hoc networks", in Proc. of *the 4th* Annual Conference on Wireless Demand Network Systems and Services, 2007.

28. X. Lin, X. Sun, P.-H. Ho and X. Shen, "GSIS: A Secure and Privacy- Preserving Protocol for Vehicular Communications",*IEEE Transactions on Vehicular Technology*, vol. 56, no. 6, pp. 3442–3456, 2007.

29. J. Freudiger, M. Raya, M. Felegyhazi, "Mix-zones for location privacy in vehicular networks", in Proc. of *ACM WiN-ITS* '07,2007.

30. J. Freudiger, R. Shokri, and J.-P. Hubaux, "On the optimal placement of mix zones", in Proc. of *the 9th* International Symposium on Privacy Enhancing Technologies, Springer-Verlag, 2009, 216–234.

31. M. Gerlach, "Assessing and improving privacy in VANET", in Proc. of *Fourth Workshop on Embedded Security in Cars (ESCAR)*, November 2006.

32. Risk-Based Systems Security Engineering: Stopping Attacks with intention, *IEEE Security $$ Privacy*, 2004, 59–62.

33. William E. Vesely, N. H. Roberts. *Fault tree handbook*,1981.

34. R. A. Kemmerer, "Covert flow trees: a visual approach to analyzing covert storage channels", *IEEE Transaction on Software Engineering*, no. 17, vol. 11, pp. 1166–1185, 1991.

35. H. Zhu, X. Lin, R. Lu, Y. Fan, and X. Shen, "SMART: A Secure Multi- Layer Credit based Incentive Scheme for Delay-Tolerant Networks", *IEEE Trans. on Vehicular Technology*, vol. 58, no. 8, pp. 4628–4639, 2009.

36. H. Zhu, X. Lin, R. Lu, P.H. Ho, and X. Shen, "SLAB: Secure Localized Authentication and Billing Scheme for Wireless Mesh Networks", *IEEE Trans. on Wireless Communications*, vol. 17, no. 10, Oct. 2008.

37. R. Lu, X. Lin, and X. Shen, "SPRING: A Social-based Privacypreserving Packet Forwarding Protocol for Vehicular Delay Tolerant Networks", Proc. *IEEE INFOCOM'10*, San Diego, California, USA, March 14–19, 2010.

38. R. Lu, X. Lin, H. Zhu, P.H. Ho and X. Shen, "ECPP: Efficient Conditional Privacy Preservation Protocol for Secure Vehicular Communications", Proc. *IEEE INFOCOM'08*, Phoenix, AZ, USA, April 14- 18, 2008.

39. R. Lu, X. Lin, H. Zhu, and X. Shen, "SPARK: A New VANET-based Smart Parking Scheme for Large Parking Lots", *Proc. IEEE INFOCOM'09*, Rio de Janeiro, Brazil, April 19–25, 2009.

40. H. Zhu, X. Lin, R. Lu, X. Shen, D. Xing and Z. Cao, "An Opportunistic Batch Bundle Authentication Scheme for Energy Constrained DTNs", *The 29th IEEE International Conference on Computer Communications (INFOCOM 2010)*, San Diego, California, USA, March 14–19, 2010.

41. C. Zhang, R. Lu, X. Lin, P.-H. Ho and X. Shen, "An Efficient Identitybased Batch Verification Scheme for Vehicular Sensor Networks", *The 27th IEEE International Conference on Computer Communications (INFOCOM 2008)*, Phoenix, Arizona, USA, April 15–17, 2008.

42. Han Q, Du S, Ren D, Zhu H (2010) "SAS: a secure data aggregation scheme in vehicular sensing networks", In: *International Conference on Communications (IEEE ICC10)*, Cape Town, South Africa, 23–27 May 2010.

43. M. Gruteser and D. Grunwald. Anonymous usage of location-based services through spatial and temporal cloaking. In MobiSys03. ACM, 2003.

44. B. Hoh, M. Gruteser, H. Xiong, and A. Alrabady. Preserving privacy in gps traces via uncertainty-aware path cloaking. In CCS07. ACM, 2007.

45. J. Freudiger, R. Shokri, and J.-P. Hubaux. On the optimal placement of mix zones. In Privacy Enhancing Technologies. Springer, 2009.

46. T. Xu and Y. Cai. Feeling-based location privacy protection for location-based services. In CCS09. ACM, 2009.

47. R.Shokri, G.Theodorakopoulos, C.Troncoso, J.-P.Hubaux, and J.-Y. Le Boudec. Protecting location privacy: Optimal strategy against localization attacks. In CCS12. ACM, 2012.

48. C. A. Ardagna, M. Cremonini, S. De Capitani di Vimercati, and P. Samarati. An obfuscation-based approach for protect- ing location privacy. Dependable and Secure Computing, IEEE Transactions on, 8(1):13–27, 2011.

49. S. Du, X. Li, J. Du and H. Zhu, An Attack and Defence Game for Security and Privacy in Vehicular Ad Hoc Networks, Peer-to-Peer Networking and Applications, Special Issue on Machine to Machine Communications, 2012.

50. S. Du, H. Zhu, X. Li, K. Ota, M. Dong, MixZone in Motion: Achieving Dynamically Co-operative Location Privacy Protection in Delay-tolerant Networks, IEEE Trans. on Vehicular Technology, to appear, 2013.

51. M. Li, K. Sampigethaya, L. Huang and R. Poovendran, "Swing & Swap: User-Centric Approaches Towards Maximizing Location privacy," in Proc. of ACM WPES06, 2006.

52. M. Gruteser and D. Grunwald, "Enhancing location privacy in wireless LAN through disposable interface identi?ers: a quantitative analysis" Mob. Netw. Appl., 2005.

53. B. Greenstein, D. McCoy, J. Pang, T. Kohno, S. Seshan, and D. Wetherall, "Improving wireless privacy with an identi?er-free link layer protocol," in Proc. of ACM Mobisys, 2008.

54. Boris Danev, Heinrich Luecken, Srdjan Capkun and Karim El Defrawy, "Attacks on Physical-layer Identi?cation," in Proc. of ACM WISEC10, 2010.

Index

S. Du, H. Zhu, *Security Assessment in Vehicular Networks,* SpringerBriefs in Computer
Science, DOI 10.1007/978-1-4614-9357-0, © The Author(s) 2013